General Editor

Dr W.J.A. Payne
Consultant in tropical livestock production

Pig Production in the Tropics

J.A. Eusebio Ph.D.

Professor of Animal Science
Former Director of Research
University of the Philippines at Los Baños

Longman Scientific & Technical
Longman Group UK Limited,
Longman House, Burnt Mill, Harlow,
Essex CM20 2JE, England
and Associated Companies throughout the world.

First published 1980
Fourth impression 1988

Set in Times

Produced by Longman Group (FE) Ltd
Printed in Hong Kong

ISBN 0 582 60617 9

British Library Cataloguing Publication Data
Eusebio, J A
Pig production in the Tropics
(Intermediate Tropical Agriculture Series)
1. Swine – Tropics
I. Title II. Series
636.4'08 SF396.T76

Other titles in the Intermediate Tropical Agriculture Series

Already published:

H. T. B. Hall, *Diseases and Parasites of Livestock in the Tropics (Second edition)*
Describes the causes, symptoms, treatment and control of the main diseases of livestock in the Tropics.

J. C. Abbott and J. P. Makeham, *Agricultural Economics and Marketing in the Tropics*
Describes the inter-relations of agriculture, farm management, marketing and their economics, as they occur in the Tropics.

C. N. Williams, W. Y. Chew and J. H. Rajaratnam, *Tree and Field Crops of the Wetter Regions of the Tropics*
Details are supplied of the botany, climatic and soil requirements, cultivation and management, harvesting and, where appropriate, processing of a large number of crops.

M. E. Adams, *Agricultural Extension in Developing Countries*
Explains the background and practicalities of extension work in the developing world.

H. F. Heady and E. B. Heady, *Range and Wildlife Management in the Tropics*
Covers all aspects of rangeland from planting and maintenance to cultural considerations.

D. V. Coy, *Accounting and Finance for Managers in Tropical Agriculture*
A useful guide to modern accounting practice for both students of agriculture and farm managers.

C. Devendra and G. B. McLeroy, *Goat and Sheep Production in the Tropics*
Provides comprehensive coverage of how to rear and maintain healthy, productive goats and sheep in the Tropics. Includes sections on breeds, nutrition, reproduction, health and breed improvement.

D. S. Hill and J. M. Waller, *Pests and Diseases of Tropical Crops. Volume 1: Principles and Methods of Control*
A comprehensive coverage of chemical control including methods of application together with information on biological and integrated control methods.

E. Heath and S. Olusanya, *Anatomy and Physiology of Tropical Livestock*
Based on body systems, this covers a wide range of agricultural animals including less familiar livestock.

D. Gibbon and A. Pain, *Crops of the Drier Regions of the Tropics*
Part 1 deals with basic ecological principles while Part 2 is a survey of different crops. Problems and potential for future development are considered in the final section.

Titles in preparation:

D. S. Hill and J. M. Waller, *Pests and Diseases of Tropical Crops. Volume 2: Field Handbook*

R. L. Humphreys, *Tropical Pastures and Fodder Crops (Second edition)*

D. H. Hill, *Cattle and Buffalo Meat Production in the Tropics*

Contents

Preface

This book is suited for students taking basic courses in swine husbandry at intermediate level. Particular care has been taken to make the text simple but sufficiently comprehensive and it is hoped that those who wish to engage in pig raising will also find this book relevant and helpful.

A large number of books have been written on pig production but a great many of these are only suited to temperate countries. Humid tropical conditions demand a different set of requirements in the manner of raising pigs. This is because the environment in general, e.g. ambient temperature, humidity, etc., exert a modulating influence on the animals' performance particularly if the breed is not native to the Tropics.

Whilst writing the book I have drawn upon my own experience in a variety of tropical countries and on the relevant literature. Also, practical applications in nutrition and integrated farming systems have been based on recent experimental researches I have conducted at the University of the Philippines.

The reader of this book should be able to acquire basic knowledge and learn the usual skills any husbandry worker should know. Emphasis has been placed on the use of cheap and locally available materials.

I am grateful to Evelina Eusebio for helping to secure data, photographs and art work. To my colleagues in the animal science profession who have shared their valuable knowledge and experiences, my sincere gratitude. Finally, my wife, Jo, deserves a lot more than gratitude for her patience, assistance and understanding during the writing and publishing of this book.

Acknowledgements

The publishers are grateful to the following for permission to reproduce photographs in the text:
Rev. A. J. Arkell for fig. 7.1c; Beyer, W. Germany for figs 5.5a and 5.7; Biophoto Associates for figs 5.6b and 5.6c; Bureau of Animal Industry, United States Department of Agriculture for figs 5.3 and 6.3; Chester White Swine Record Association for fig. 1.2; Bruce Coleman Ltd. for fig. 5.6a; J. A. Eusebio for figs 1.9, 2.5a, 2.5b, 2.6, 2.7 and 5.1; Federal Ministry of Information, Lagos for fig. 7.1a; Rob Francis for figs 1.16 and 3.6; Desmond Hill for figs 1.15, 6.1 and 7.1b; Alan Hutchison Library for figs 1.13 and 1.14; National Pig Breeders Association for figs 1.1, 1.6a, 1.7, 1.10. 5.4a, 6.2 and 6.6; National University of Colombia for fig. 1 (introduction); W. J. A. Payne for figs 2 (introduction), 1.5, 1.11 and 7.13; Pictor International for figs 1.12 and 1.17; Plum Island Animal Disease Laboratory, New York for figs 5.2a and 5.2b; Poland China Record Association for fig. 1.8; Vernon G. Pursel for figs 1.3, 1.4 and 1.6b; The Slide Centre for figs 2.3, 6.4, 6.5, 6.7, 6.8, 6.10a, 6.10b and 6.10c; C. James Webb for fig. 5.8; Zambian Information Services for fig. 9.3. The publishers would like to thank Pictor International for supplying the photograph of the sway-back pig from Bali on the cover.

Glossary

abortion Premature expulsion of a non-living foetus.

adhesion Adhering; attachment or sticking fast to.

aerobic Living or functioning in the presence of air or free oxygen.

agglutination test Method of determining rate of clumping together of cells or suspended particles.

algae Group of plants containing chlorophyll but without stems, roots, leaves; single-celled algae form scum on rivers, ponds and lakes, and large forms are seaweeds.

ambient temperature Temperature all around; e.g. in a room or any surrounding environment.

anaerobic Living or functioning in the absence of oxygen or air.

analeptic Has ability to restore strength; reinvigorating; medicine restoring vigour.

anorexia Lack or loss of appetite for food.

anthelmintic De-wormer; a chemical destructive to intestinal worms.

antibody Substance produced in the body in response to invasion by a foreign body or to antagonistic substances such as those produced by micro-organisms.

antioxidant Substance that delays or prevents oxidation, e.g. alpha-tocopherol and ascorbic acid, BHT and BHA.

anus Terminal portion of the rectum; outlet of the alimentary canal.

apocrine sweat glands Glands producing sweat by the breakdown of part of the active cells of the gland.

arthropod parasite Invertebrate animal with jointed legs and a segmented body which lives on or within another organism from which it derives subsistence.

ascarid Roundworm which is a parasite of the pig.

atlas joint Bone joint at the first vertebra of the neck.

barrow Male pig castrated before attaining sexual maturity.

bile duct Passageway for bile after it has been produced by the liver.

Boston butts Upper halves of the pork shoulder.

bronchi Branches of the trachea that are the air passages supplying the right and left lungs.

bronchioles Fine tubes branching repeatedly from each bronchus in the lung.

brood sow Mother pig with her young piglets.

buffer Agent that resists marked changes in acidity (pH, hydrogen ion concentration); plays a vital role in the regulation of acid-base balance in the body.

castrate To remove the testicles of the pig.

causative agent An organism or thing responsible for an action or result such as bacteria or parasites which cause disease.

chine A cut of meat containing part of the backbone.

clear plate X-ray plates indicating absence of pulmonary diseases.

cleaver A butcher's tool with a heavy blade and a short handle, used for cutting through meat or bone.

clinical manifestation Physical signs in the pig generally associated with sickness or disease.

colostrum First milk secreted after sow gives birth; usually higher in protein and antibodies needed for immunity in new-born pigs.

conception rate The number of times a sow becomes pregnant; or the number of sows in the herd that become pregnant.

confinement feeding Giving feeds inside pig pen or house instead of grazing in pasture.

constipation Difficulty of bowel movement because of sluggish colon causing retention of faeces.

convulsive attack Sudden violent involuntary contraction of muscles.

corpus luteum Reddish yellow mass of endocrine tissue which develops in the ovary from a ruptured Graafian follicle.

cyst Sac, especially one containing a liquid or semisolid.

dam A female parent pig.

dehydration Excessive loss of water from food, tissue or body.

diagnosis Finding out what disease an animal has by examination and careful study of symptoms.

diaphragm A dividing partition of muscles and tendons separating the cavity of the chest from the cavity of the abdomen.

diarrhoea Too many and too loose bowel movements; too frequent passage of faeces.

diffuse Spread out to cover larger surface or space; disperse.

disaccharides Sugars containing two monosaccharides joined together. Examples are lactose, maltose, sucrose.

dropsy Abnormal accumulation of fluid in a body cavity or cellular tissue.

ejaculate To eject or discharge, especially semen.

emaciated Made thin from loss of flesh due to starvation or illness.

embryonate To develop into an embryo; in the early stages of development.

embryonic mortality Death during the early stages of development of an organism.

emulsify To make into an emulsion, a milk-like liquid containing very tiny drops of fat.

encapsulated Enclosed in or as if in a capsule.

endocrine glands Ductless glands producing hormones (internal secretions) discharged directly into the blood stream.

environmental stress Pressure, force or strain on health from conditions in the environment.

enzyme Organic catalyst produced by living cells; these control most of the chemical reactions and energy transformations in both plants and animals.

epididymis A convoluted tube attached to the upper part of each testicle, in which sperms are stored.

evisceration Removal of vital parts such as internal organs, especially intestines of the pig.

exocrine sweat gland Gland producing a secretion which is poured on to the external surface of the body through ducts.

exudate A substance given off or oozing out of the wall of a tissue.

eye of lean meat *Longissimus dorsi* muscle.

farrowing Process of giving birth to a litter of pigs by a mother sow.

feed efficiency Amount of feed required for an animal to make a unit gain in weight.

ferment Undergo chemical change becoming sour or alcoholic and giving off gas or bubbles.

fibrin Protein substance which serves as the essential network in which blood cells are enmeshed to form a clot.

filtrable virus A disease agent so small that it can pass through porcelain filters; cf. non-filtrable viruses which cannot pass through.

fistula Artificial opening or tube-like passage in the body.

foetus Embryo; the unborn young of an animal while in the uterus.

follicle Any small sac, cavity or gland for excretion, as in the ovarian or hair follicles.

foreshank The lower part of the front leg of the pig.

gambrel Hard wood or steel device strong enough to hold a desired weight.

gastric gland Gland in the stomach which produces secretions containing enzymes.

gene Unit on the chromosomes which carries hereditary characteristics; a specific portion of DNA.

genitalia The external parts of the reproductive system or the sex organs.

gestation period Period of pregnancy: in pigs, usually 3·5–4 months.

gilt Young female pig approximately 6–8 months of age.

gonadotrophic hormone Hormone secreted by the pituitary gland which causes the ovary to secrete the female sex hormones.

haemorrhage Bleeding; profuse escape of blood from a part of the body.

helminthic parasite Worm or worm-like animal in the intestine, e.g. tapeworm, hookworm or roundworm.

heritability A feature which has this is capable of being inherited or of being passed on by inheritance.

heterozygous A genetic term meaning that a given genetic factor possesses both the dominant and the recessive characters of the gene-pair. One character, in any pair will be derived from the sperm and one from the egg.

hock Joint bending backwards in the hind leg of the pig.

hog-tied All four feet or hands tied to keep in place.

homozygous Containing identical genes in the two corresponding loci of a pair of chromosomes.

hormone Substance produced by the endocrine or ductless glands in the body (such as the adrenal glands or the pituitary) and carried to another organ or tissue where it has a specific regulatory effect.

hybrid vigour Superiority of crossbred progeny over their parents.

hypothalamus Area in the brain responsible for the maintenance of body temperature, blood pressure, water regulation and the control of appetite and other basic functions of life.

immunisation Rendering immune from infections and

disease by vaccinations.

implantation Insertion of a tissue or part of the body in another part of the body.

incubation Period of development under controlled conditions such as uniform temperature and humidity.

infusion Liquid extract that results from soaking a substance in water.

insecticide A substance to kill insects.

inventory A listing of animals and materials in stock.

jowl Lower jaw of the pig with the chin and cheeks.

larvae Usually applied to insects from time of hatching until transformation into pupa.

leaf fat A heavy layer of fat around the kidneys of a pig; it is used for making lard.

lesion Any injury or cut; any loss of function of a part of the body.

litter Set of young pigs born at one time to the mother pig.

loin Lower part of the back on either side of the backbone between the hips and the ribs.

lymph Fluid obtained from lymphatic ducts; one of the circulating fluids in the body important for fat absorption.

mammary gland Milk-secreting gland in the breast of the sow.

mesenteric fat Fat that enfolds some internal organ and attaches it to the body wall or to another organ.

metabolic activity Chemical changes of nutrients and substances in the body in order to produce energy and new materials for the body.

methane Odourless, colourless, inflammable gas produced from anaerobic respiration of pig manure and used as a fuel.

monosaccharides Group of carbohydrates which are simple sugars, e.g. glucose, ribose, fructose and galactose.

morbidity Diseased or unhealthy condition; prevalence of a disease.

mortality rate The ratio of number of deaths to the total pig population.

mucus Viscous substance secreted by glands in mucous membrane.

necrotic Dead cells or decayed tissue or bone surrounded by healthy tissues and still part of the living body.

nephrosis Condition of degeneration and disintegration of the kidney without signs of inflammation.

nutritional anaemia Reduction in the number or size of red blood cells or of the quantity of haemoglobin due to a deficiency of nutrients necessary in the formation of blood. Iron, protein, folic acid, vitamin B_{12}, cobalt and copper are essential nutrients.

oesophagus Hollow muscular tube extending from the pharynx to the stomach.

omnivorous Eats any type of food.

ovary Female sex gland which produces eggs (ova) and secretes sex hormones.

ovum Egg; a female cell which after fertilisation can develop into a new animal.

parturition Process of giving birth.

pathogenicity The capability of developing or producing a disease.

pedigree Recorded line of descent, especially of a purebred pig.

operi-renal fat Layers of fat around the kidney.

pharynx Muscular and membranous tube extending from the oral cavity to the oesophagus.

photosynthesis Production of carbohydrates from water and carbon dioxide using energy from the sun.

picnic (Pertaining to a pork cut) – the lower half of the pork shoulder.

pigmentation Coloration; deposition of pigment in cells.

placenta Organ within the uterus which serves for the respiration and nutrition of the growing foetus.

pollards A grade of wheat bran used as a pig feed.

polypeptide Protein derivative; composed of shorter chains of aminoacids than protein.

portal blood circulatory system Blood transport system by which nutrients are carried to the liver via capillaries and veins after absorption from the intestines.

posterior paralysis Loss or impairment of motor function at the rear or back portion of the body.

prehensile Adapted for seizing or grasping.

pre-starter ration Diet given to baby pigs before they are fully weaned from the mother sow.

prolific Producing many young.

prostration Exhaustion; to lie on the ground due to physical weakness.

protozoa Group of mostly microscopic, single-celled animals living chiefly in water, but sometimes parasitic.

pyloric glands Glands located along the opening between the stomach and the duodenum (first part of the small intestine).

quarantine Keeping an animal having an infectious

disease away from others usually for a period of 40 days.

rancid Applied to spoiled fats and oils having a bad smell, odour or taste.

renal pelvis Basin-like cavity containing the kidneys located in the lower part of the pig's body.

reproductive efficiency Number of pigs per litter or number of farrowings per unit time.

scour Watery diarrhoea in the pig.

scrotal hernia Protrusion of the testes of the male pig from the scrotum.

scrotum The external sac of skin enclosing the testes.

sedimentation The accumulation or deposition of insoluble material in a liquid at the bottom of the container.

semen Fluid secreted by the male reproductive organs which contains the spermatozoa.

septicaemia Condition caused by the presence of pathogenic bacteria and their toxins in the blood. It is accompanied by chills, intermittent fever, excessive perspiration and weakness.

sire The male parent of an animal.

spermatogenesis Process of production of sperms or male germ cells in the reproductive organs.

sterility Incapacity or inability to produce others of its kind.

stillborn Dead at birth.

symbiosis Intimate living together of two different kinds of organisms or groups dependent on each other.

syndrome An aggregate or set of concurrent symptoms together indicating the presence and nature of a disease.

thromboplastin A complex substance found in the blood and other animal tissses which reacts with calcium ions to give prothrombin in the process of blood clotting.

thumps A lung disease in pigs, caused by infection with the larvae of a roundworm.

toxaemia Condition in which the blood contains toxic or poisonous substances either produced by the body or by micro-organisms.

trachea Wind pipe of the respiratory system; tube with supporting rings of cartilage extending from the larynx to the bronchi.

trotter The foot of a pig used as food.

tunicia vaginalis Inner membranous layer of the testes of the male pig.

udder A large, pendulous, milk-secreting gland provided with nipples or teats for the baby pigs to suck.

urate A salt of uric acid.

ureter Long, narrow tube that conveys urine from the pelvis of the kidney to the bladder.

uterine infection Disease of the uterus caused by infection with bacteria or viruses.

vaccine Suspension of killed or weakened micro-organisms, or of products derived from them, that produce immunisation against a disease on introduction to the body.

vector A carrier, usually an arthropod, that transmits disease-producing foreign agents from one infected animal to another.

vent Small opening or outlet; passage to release gases.

visceral organs Commonly, the digestive organs within the abdominal cavity. Also the internal organs of the body such as the heart, lungs, liver, intestine, etc.

vomiting Throwing up or ejecting the content of the stomach through the mouth.

wallow To roll about in mud or dirty water.

weaning Transferring the young pigs from dependence on the mother's milk to another form of feed.

Introduction

Pig production is big business in Europe and North America. The use of improved breeds and progress in feeding and management practices, including disease control measures, have been important factors in developing the industry in these regions of the world.

In the poorer countries of the world, however, especially in the humid tropical regions, with the exception of some countries in Southeast Asia, improvement in pig production has been quite slow compared to improvements in poultry production. In many tropical countries, there are still a large number of domesticated native pigs which are scavengers or are raised in the backyard to depend on kitchen left-overs and farm wastes. Nevertheless, in spite of the low level of management, pigs have continued to provide a significant amount of animal protein thus improving the diet of local people. Table 1 on p. 2 shows the sources of animal protein in a number of regions and on a world-wide basis.

The productivity levels, in terms of meat yield of cattle and pigs, differ from country to country. Meat productivity data from 1956 to 1968 has shown that pigs can yield on the average, 4 to 5 times more meat than cattle per tonne of liveweight (Table 2, p. 2). However, this ratio varies from country to country and region to region. In South America, pigs yield only twice as much meat as cattle. In Asia and Africa, the meat yields per tonne of liveweight were 25 and 8 times as much, respectively.

This variation in the level of meat productivity of cattle and pigs results from the use of different breeds and the level of management practices, including disease and parasite control measures.

An FAO report on the world pig population indicates that there was an average increase of 23 per cent in the pig population of the Tropics during the period from 1969 to 1974. There were significant increases in pig population in the tropical regions of America and Asia with 27 and 18 per cent increases, respectively (Table 3, p. 3). In 1974 the pig population in China represented 35·8 per cent of the total world pig population; the Chinese are traditionally a pork-eating people. In the tropical countries of Africa, in some of which there are large Muslim communities, the pig population represented only 0·8 per cent of the world total. Pig production in this region is also restricted by long dry seasons and the unavailability of cheap grains and their by-products as feeds.

Advantages of pig raising

Efficient conversion of animal feed to human food From the standpoint of the efficient production of meat the pig is superior to beef cattle, goats or sheep when the feed provided is of a high quality. When the feed is of lower quality, e.g. rice bran, grass, hay, etc., the pig is not as efficient as ruminant livestock.

Financially rewarding The capital invested can be realised and returned in a relatively short period. It takes a 6 to 7 months feeding period to raise a weanling pig to a market weight of 90 kg, under average feeding and management conditions in a humid tropical climate.

Omnivores Pigs are voracious eaters of farm crop waste products and offals of livestock and poultry which are converted efficiently into pork, thus yielding edible protein and a food of high calorific value. Kitchen left-overs or restaurant refuse can also be converted efficiently into meat. Farm grain that has been damaged by rain or fire can be eaten by mature pigs. Unlike cattle, pigs may even be fed on mouldy grain. It has been reported that cattle fed mouldy grain have shown a high death rate, while pigs fed with the same feed did not show the same adverse effects.

Tolerant to a wide variety of feeds Pigs of all classes, except young piglets, can tolerate all kinds of feeds, even to some extent low quality, highly fibrous foods. It has been a practice in developed and underdeveloped countries engaged in pig production to feed pregnant sows with freshly cut forage or corn soilage mixed with a small amount of protein feed concentrate. This reduces energy intake economically and improves the sow's reproductive efficiency.

Large litters A sow can easily produce a litter of 8 to 12 pigs after a relatively short gestation period of 112–120 days. In many underdeveloped tropical countries in which pork consumption is not restricted by religious customs and beliefs, pig raising can be a means of filling the gap between the production of protein food and the rapid rate of population increase. Pork is not only rich in protein but is also a good source of energy.

Live in small areas Pigs need only a small space in which to grow. Unlike beef and dairy cattle which usually require at least 1 ha of natural pasture per head, pigs can be raised on a small area either in close confinement

Table 1 Comparative contribution of different types of meat in the supply of animal protein. (*Source:* Kroeske, D. (1971). *Fifth FAO Reg. Conf. on Anim. Prod. and Health in the Far East*, Malaysia.)

Region	Animal protein in meat (g) per head per day	Percentage contribution to total				
		Beef	Mutton	Pork	Poultry	Other meat
Far East	3·3	22	7	45	15	11
Africa	5·5	50	16	3	7	24
Near East	5·7	36	42	0	4	18
Latin America	12·9	64	4	13	6	13
Europe	18·3	39	6	34	9	12
North America	36·9	50	2	22	20	6
World	9·4	42	5	29	13	11

Table 2 Meat productivity of cattle and pigs, expressed as meat yield in kg per year per tonne liveweight. It is assumed that 5 head of cattle, including young and adult animals and 20 head of pigs (including all ages) weigh, on average, 1 tonne. (*Source:* Kroeske, D. (1971). *Fifth FAO Reg. Conf. on Anim. Prod. and Health in the Far East*, Malaysia.)

Region	1956		1966		1967		1968	
	Cattle	Pigs	Cattle	Pigs	Cattle	Pigs	Cattle	Pigs
Europe	250	1 560	310	1 780	380	1 780	335	1 780
USSR	160	1 160	200	1 140	225	1 160	240	1 200
North America	190	1 820	335	2 140	335	2 120	345	2 100
South America	145	340	135	300	140	320	145	300
Asia	25	520	30	800	30	820	30	840
Mainland China	165	960	165	920	170	900	170	880
Africa	65	580	80	580	75	560	75	580
Pacific	220	1 320	225	1 360	235	1 380	230	1 360
World	140	1 120	160	1 100	165	1 100	170	1 120

Table 3 World pig population. (*Source:* FAO (1975). *Monthly Bulletin of Agricultural Economics and Statistics*, **24** (2), 21–22.)

Region	1968/69		1973/74		Percentage increase or decrease
	Number (thousand)	As percentage of world population	Number (thousand)	As percentage of world population	
Europe	124 167	22·00	154 934	23·00	+ 25·0
USSR	49 047	8·50	70 032	10·00	+ 43·0
Africa	6 337	1·00	7 183	1·00	+ 13·0
Tropics	4 855	0·85	5 568	0·80	+ 14·0
Others	1 482	0·26	1 615	0·20	+ 8·9
America	126 988	22·00	137 987	20·00	+ 8·6
Tropics	49 684	8·60	63 355	9·00	+ 27·0
Others	77 304	13·50	74 632	11·00	– 3·4
Asia	261 872	45·70	292 828	43·80	+ 11·8
Tropics	33 948	5·90	40 347	6·00	+ 18·6
Others	9 913	1·70	12 907	1·90	+ 30·0
China	218 011	38·00	239 574	35·80	+ 9·8
Pacific	4 138	0·70	4 480	0·67	+ 8·2
Tropics	160	0·03	176	0·03	+ 1·0
Others	3 978	0·69	4 304	0·60	+ 8·0
World	572 549		667 444		+ 16·5
Tropics	88 647	15·00	109 446	16·00	+ 23·0
Others	483 902	85·00	557 998	84·00	+ 15·3

within a building or on a small area of pasture. Even in suburban areas, pigs are raised either in the backyard with 1–3 head per family or on a medium production scale with an average of 20–50 head per family. A mature sow or active boar requires only 4–5 m^2 of living space.

High percentage of useful products per pig The results of studies conducted at the University of the Philippines at Los Baños, indicated that a finished pig can yield 70–75 per cent of dressed carcass. In addition, the entrails and blood can be used for sausages as practised in South America and elsewhere. The skin may be cooked to produce crackling which sells at a high price in supermarkets. It is said that the packers make money out of everything except the pig's squeal. The hairs are made into brushes and the hoofs into glue. The bone can be ground into bone meal for livestock feed. Table 4 shows average dressing percentages i.e.

$$\frac{\text{kg dressed carcass}}{\text{kg liveweight}} \times 100$$

of the different farm animals including poultry.

Table 4 Average dressing percentage of different farm animals. (*Source:* Department of Animal Husbandry, University of the Philippines, College of Agriculture, 1970.)

Animals	Dressing percentage
Poultry (broiler)	64·3
Hog (70–90 kg liveweight)	75·6
Steer (Batangas/Philippines breed), 170–310 kg liveweight)	52·2
Sheep	50·0
Goat	44·0
Rabbit	48·1
Horse	49·1

Improve soil fertility Like other farm animals, pigs contribute a considerable amount of fertilising ingredients to the soil through their manure. A mature pig can produce 600–730 kg of manure annually. The nitrogen content of fresh pig manure ranges from 0·55–0·6 per cent; phosphate(v) content, 0·5 per cent; and potassium content, 0·4 per cent. With the present inorganic fertiliser shortage, organic fertiliser from the manure of farm animals will help supply some of the soil nutrients required by plants, especially vegetables. In some sugar-producing countries such as the Philippines and Colombia, pig manures are drained from big piggeries into irrigation canals which irrigate the sugar cane fields (Fig. 1).

fig. 1 Pig waste drainage canal to a sugar plantation

Profitable for small farmers Many poor families in rural areas in the less developed tropical countries of Southeast Asia and South America, raise one or two pigs as their 'savings banks'. These families schedule the raising of their pigs so that they can sell them at the beginning of the school year when they have to pay the school fees for their children. The poor families in the suburban or rural areas find raising one or two pigs in their backyard profitable since the feed used usually comes from left-overs in the kitchen and from the farm.

Disadvantages of pig raising

Pork is not an internationally accepted food Consumption of pork is forbidden by certain religions and pig production is not encouraged in Muslim countries in tropical Africa and Asia. Whereas milk or poultry products are generally acceptable.

Suburban pollution Because pigs have a single stomach and require only a small space, they are usually raised on a backyard scale in suburban communities where kitchen left-overs are plentiful. As a consequence, the surroundings of these communities may swarm with flies which could cause excessive pollution. To minimise the problem of pollution, pig manure and other waste can be utilised in the production of methane as practised in Taiwan, the Philippines and other countries.

Susceptibility to parasites and diseases In many underdeveloped countries in the Tropics, there are still a large number of pig producers who raise pigs in a traditional scavenging system. Scavenger pigs are not only particularly susceptible to parasites and diseases but they are also carriers of disease, e.g. swine cholera and plague. In humid tropical countries the climatic environment encourages the development and spread of parasites and disease throughout the year. In temperate countries seasonal climatic changes reduce the overall incidence of parasites and diseases.

Competition with people for food grains Increases in the price of cereals during the last decade in some countries may have been partly due to an increase in the number of large-scale commercial pig farms. Pigs are more efficient converters of maize or sorghum into edible meat than ruminants. In some countries, large-scale pig producers taking advantage of this have used large quantities of grain, particularly as there is a fast turnover of the capital invested in pig production. As a result, maize and other grains are sometimes hoarded because they command better prices as pig feeds than as human food.

fig. 2 (a) Tethered pig

Pig raising systems

There are four major systems of pig raising in many of the less developed tropical countries in Asia, South America and Africa.

Scavenging pigs

The pigs are let loose day and night on a self-supporting feeding management system. In this system, native or upgraded strains of pigs are predominantly used because they are more tolerant of low-quality feeds. Also, scavenging pigs can be resistant to some parasites such as ascaris and lungworm. The pigs are generally marketed according to the financial needs of the owner rather than with regard to their weight. The quality of meat produced from scavenger pigs is inferior with regard to deposition of meat. Fig. 2 (b) shows a scavenging pig.

Modern breeds of pigs, or any crossbreds survive with difficulty under this system of raising. This is because they are susceptible to environmental stress. The feeding requirements for the modern large breeds of pigs are more exacting than those for the small local breeds used as scavengers.

(b) Scavenging pig

Backyard pig raising

This system of swine raising is found in most villages and suburban areas where a family may keep up to 3 pigs in the backyard. The pigs are kept in an elevated bamboo or other simply constructed sty or are tethered with a rope around the heartgirth in the yard or elsewhere.

The reproductive performance of the sows is poor because they receive minimum care and feeding management. The litter size at farrowing averages 8 to 10 pigs of which less than half survive to weaning at 2 months of age. Sexual maturity of local pigs or upgraded strains is attained at 4 to 5 months of age with average body weights of 40 kg. In some countries, e.g. the Philippines, Thailand and Taiwan in Southeast Asia, and Colombia, Panama and Mexico in the Americas, crosses of modern breeds of pigs are also kept under the backyard system.

Pigs managed within this system depend on kitchen left-overs with occasional supplements of rice bran or maize by-products. Pigs are marketed indiscriminately, the major consideration being the urgent financial needs of the family.

Medium-sized pig units

In this system of production the size of the herd ranges from 20 to 50 head of all ages, and the producer depends on feeding mostly commercial feeds. The kitchen left-overs are obviously not sufficient to sustain the feed requirements of the herd. Farm mixing of the feeds needed for this size of operation may be uneconomic because the costs of the mixing equipment and operational mixing expenses are quite high and are not justified with such a small number of animals to feed.

Most medium-scale pig producers are dependent on middlemen for the marketing of their produce. The middlemen may also be medium-scale pig producers.

As a consequence of the above factors more often than not, medium-scale pig farming operations are less profitable than either backyard or large commercial operations.

Large-scale commercial pig production

It is a common practice for a large-scale commercial pig farm to integrate its operation from feed-grain production through feed manufacturing, pork production and processing of products to marketing. Large-scale operations may involve a total sow herd of from 50 to as many as 1 000 head. The total pig population of all ages could range from 600 to 12 000 head at any one time.

Future outlook for pig production

Valuable information on livestock and poultry farming in the humid Tropics, that can be used for future animal development programmes, has been gathered by some international and national organisations. There is evidence that pig production in humid tropical countries has a bright potential. It is also claimed that pigs can be successfully and economically raised in the Tropics because building costs are low and the heating facilities required for young pigs are minimal. Pigs can thus be protected against the climatic environment by the provision of good housing facilities.

The major problems in pig production in many of the less developed countries in the Tropics consist in an uncertainty of the supply of cheap feeds, and the continuous danger of diseases such as hog cholera, swine plague and African swine fever. Some of these problems can be overcome by application of the new technology and management practices which have resulted from research studies conducted in advanced temperate and tropical countries. There is also new information on the utilisation of many non-conventional feedstuffs, e.g. hydrocarbon yeast, algae, leaf proteins, etc.

This book provides basic and practical information on the raising of pigs under tropical climatic conditions. It should prove useful both for students and for workers in the field.

Part 1 Pig breeding

1 Breeds

Practically all domesticated breeds of pigs are believed to have originated from two major wild types, *Sus vittatus* and/or *S. indicus* – the wild pig of China, Japan and Southeast Asia and *S. scrofa* – the wild pig of Europe, which probably also originated from the Asiatic continent. Pigs from Asia may have been introduced into Europe by the early settlers who brought other domestic animals with them.

In the eighteenth century, Chinese and Siamese pigs were introduced into the British Isles and crossed with local breeds. The Siamese pigs were copper-coloured with black hair. This type of pig was presumed to have been brought into America during the early colonial days and later crossed with improved English breeds or with pigs from other parts of the world.

Domesticated pigs were also introduced in the Pacific Islands, the Caribbean and Australia. European-type pigs and the Indian pig known as *S. cristatus* possibly arrived in Melanesia by way of Southeast Asia.

Indigenous and introduced pigs are now being improved through breeding and selection to meet the demands of the market and to fit into specific environments. Thus the breeds selected should be able to utilise locally grown feeds to the greatest advantage.

Three types of pig have been developed to their highest efficiency in Europe and in America; the meat, bacon and lard types. Today, however, pig breeders claim that practically all present-day breeds' possess the desirable characteristics of the meat-type pig.

The following breeds, including indigenous tropical pigs, have been widely utilised either as purebreds or as crossbreds.

European and American pigs

Berkshire

The Berkshire, which is believed to originate from a cross between old English and oriental pigs, had its beginnings in south central England particularly in the counties of Berkshire, Wiltshire and Gloucestershire. The ancestry of the English pigs can be traced to *Sus scrofa*, the wild boar of European origin.

The other ancestors of the Berkshire, the oriental pigs of Chinese and Siamese origin, belonged to the species *S. vittacus* and *S. indicus*. Compared to the English pigs, the Chinese pigs were much smaller though their sows were prolific.

The modern Berkshire, a medium-sized animal, has a very distinct black haircoat with 6 white points (Fig. 1.1). The 6 points are on each of the 4 feet, on the face and on the tip of the tail. Occasionally, however, there is a splash of white on the body. The curved face is a very distinctive feature of the Berkshire.

In the Tropics, Berkshires are considered prolific enough under average farm conditions, sows averaging 8 to 9 pigs per litter.

Unlike most temperate breeds of pig the Berkshire is not particularly large at maturity, but it grows rapidly and is a good grazer. Under tropical conditions and with good feeding and management, a Berkshire pig can easily weigh 70–80 kg liveweight at about 6 months of age.

Chester White

These white pigs are claimed to have originated from the Yorkshire and the Lincolnshire, both English breeds, and the Cheshire breed of pigs that came from Jefferson County, New York.

Except for a few black or small bluish spots that sometimes appear on its skin, the Chester White is solid white in colour (Fig. 1.2). It is a large lard-type pig, hardy and a fairly good feeder.

Chester Whites farrow and raise litters similar in size to Yorkshires. The conception rate has been the best of all American or British breeds. Carcass quality is intermediate as it is shorter and contains more fat on the average than other breeds. However, carcasses produce a good size loin muscle and a high percentage of ham. The growth rate of this breed is slow compared with other breeds.

fig. 1.1 Berkshire pig

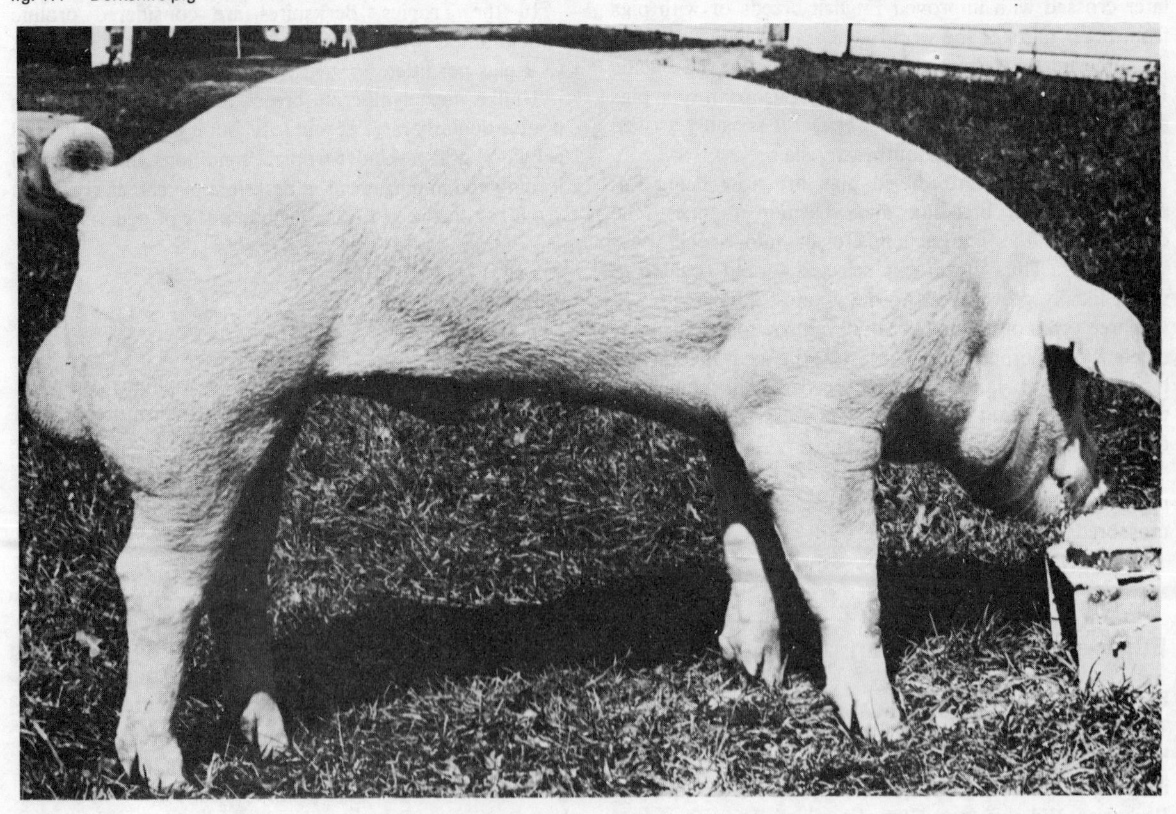

fig. 1.2 Chester White boar

Chester Whites have been criticised for their white colour which is considered a great handicap. They easily become dirty and unattractive and readily sunburn. Through selection the breeders have produced pigs with heavier coats and a thicker white skin that does not sunburn so easily.

Duroc–Jersey

The Duroc–Jersey or Duroc originated from the eastern USA. The two strains of pigs used in forming the Duroc–Jersey breed were the Jersey Red of New Jersey and the Duroc of New York. Some authorities have suggested that the Tamworth breed was also involved.

Durocs are characterised by solid colours that range from a very light golden to a dark red that approaches the colour of mahogany (Fig. 1.3). Many farmers, especially those in South America and Southeast Asia, like this coloration.

The Duroc excels all other breeds in muscle quality and probably has the lowest incidence of stress mortality. It has mothering ability and litter size is average. The carcass possesses only an average backfat thickness and percentage of desirable cuts. Lack of soundness of the front legs is a serious problem within the breed, and careful selection against this characteristic is required to prevent its introduction into the herd.

The desirable characteristics of the Duroc include the ability to adapt to varied conditions in the environment, sturdiness, and the ability to thrive well even on plain kitchen refuse. The breed's efficiency in converting feed consumed into liveweight has been responsible for its increasing popularity among pig raisers in the Tropics.

The high rate of pig mortality in the Tropics has led many pig raisers to replace other temperate breeds with the Duroc because of the latter's hardiness. In most cases, these producers cross the Duroc with other temperate breeds.

fig. 1.3 Duroc–Jersey pig

fig. 1.4 Hampshire pig

fig. 1.5 Hampshire–Florida crossbred pig

Hampshire

The Hampshire's native home was southern England, where black pigs with white belts were crossed with the Wessex Saddleback. Hampshires are black pigs with a white belt around the forequarter of the body. It is this peculiar white belt characteristic that gave the Hampshire its popularity (Fig. 1.4). Other features of the Hampshire are a long, straight face and erect ears. At maturity, the breed is medium sized. Hampshires are prolific and good nursing mothers.

Hampshires have been reported to be the leanest breed of pig in the USA. The length and the percentage of ham and loin in the carcass are excellent. Feed conversion of this breed has been found to be superior under test-station conditions. However, Hampshires show poor cleanliness characteristics under confinement conditions, and some strains have experienced some stress-adaptability problems. Growth rate of the breed has generally been average or below average. Hampshire sires are excellent for use in the final crossbreeding, especially where carcass leanness is important (Fig. 1.5).

Landrace

This famous breed of pig had its earliest beginnings in Denmark. In 1895, Large White boars were brought from England and crossed with the native pigs. After years of breeding and selection by pig progeny testing stations, sponsored and financed by the State's Research Laboratory of Animal Husbandry, Copenhagen, and the private sector and farmers of the Cooperative Bacon Factories, breeders produced a bacon pig that satisfied most requirements for a bacon type. Between $\frac{1}{16}$ to $\frac{1}{64}$ of Poland China blood was introduced into the Danish Landrace breed to establish the foundation herds of the American Landrace (Fig. 1.6 (a) and (b)).

The Landrace breed is characterised by a solid white colour, which is sometimes freckled, and 16 or 17 pairs of long ribs.

It has not achieved the popularity in tropical countries that might have been expected. When first introduced in the Philippines, the majority of commercial pigs were managed in close confinement in concrete pens. Under these conditions the breed demonstrates consistent

fig. 1.6(a) Landrace sow

fig. 1.6(b) Landrace boar

lameness. Recent trends towards improved rearing of commercial pigs in confinement and the renewed emphasis on better maternal performance has, however, increased the popularity of the breed. New introductions from Europe via Canada have also improved the performance of the breed. Although litter size, mothering ability and confinement adaptability are excellent, when the pigs are full fed high-energy diets, carcass quality deteriorates relative to many other breeds.

In the Philippines, Taiwan, Singapore and some African countries, the Landrace is one of the breeds used in local pig-breeding programmes. Many pig raisers in other countries are now crossing Landrace pigs with other temperate-type, coloured breeds.

Given proper feeding and management, Landrace will grow well in the Tropics, although they do not readily adjust to the consumption of local feedstuffs, such as rice and corn meal.

Large White or Yorkshire

The Yorkshire's earliest home was in Yorkshire, UK, and the surrounding counties. Its ancestors can be traced to the large white pigs of heavy bone and great length and a skin that had dark pigmented spots. The present Large White breed was developed in UK by selecting and crossing Yorkshire with White Leicester pigs. Large White sows are prolific giving large-sized litters.

The Yorkshire breed is much more muscular than American-developed breeds. Because they are white, Yorkshires are very susceptible to sunburn, especially at the base of their ears (Fig. 1.7).

The Yorkshire breed has proved to be one of the most superior breeds in litter size and mothering ability, and it adapts well to confinement conditions. Yorkshire sows have been superior to those of all other European and American breeds in the number of the litter raised in test studies. Growth rate of the breed is excellent, especially

fig. 1.7 Yorkshire pig

under confinement conditions. Some animals of this breed produce carcasses with excess backfat and some difficulty has been experienced with the incidence of pale, soft muscles and stress mortality in smaller framed, more muscular strains.

Farmers find the Yorkshire valuable in their breeding and upgrading programmes. These qualities plus its voracious feeding habits have encouraged pig raisers in tropical countries to use the breed in both purebred and commercial pig operations.

Poland China

An old breed of pig, the Poland China was known earlier as 'Hot Type' or 'Big Type Poland China'. The Poland China is believed to be a crossbred developed in the USA from Russian, Byfield, and Big China pigs.

The Poland China breed has a colour pattern that is similar to that of the Berkshire. This is because the Big China or the Warren hogs had been improved by previously crossing them with the Berkshire.

Like the Berkshire, it has 6 white points, one on each of the 4 feet, one on the tip of the tail, and one on the tip of the nose. Poland China pigs are generally larger than other modern breeds (Fig. 1.8).

The Poland China's growth performance has been fairly good in the Tropics, although it is susceptible to kidney worms and other parasites.

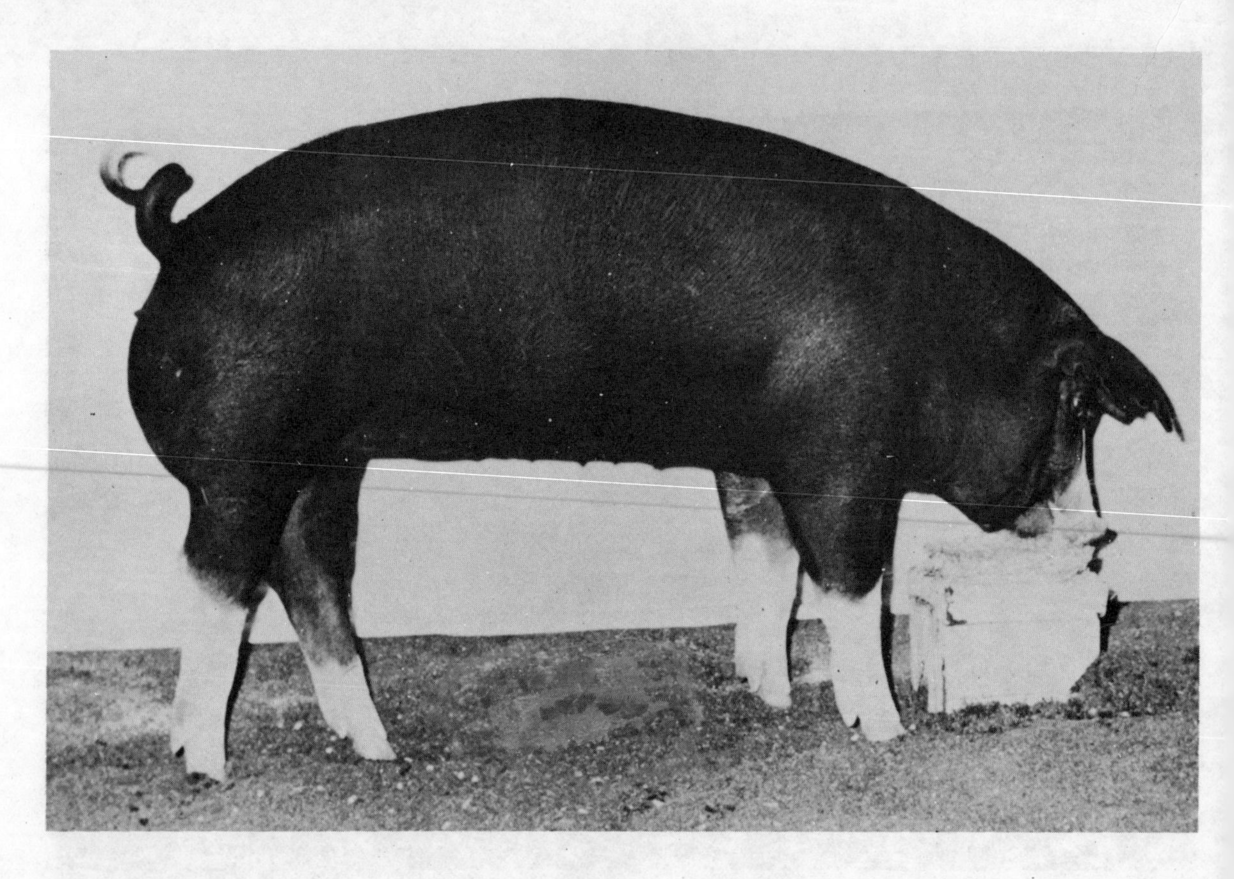

fig. 1.8 Poland China pig

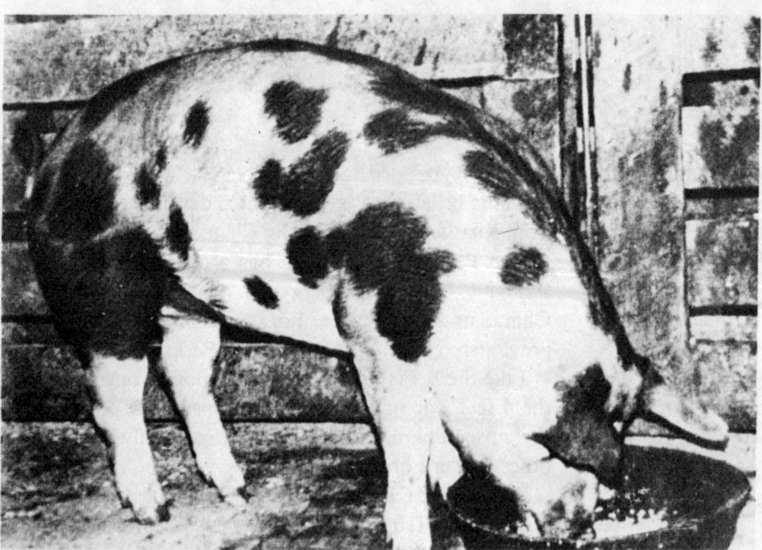

fig. 1.9 Spotted Poland China pig

Poland Chinas are efficient feed converters. With good feeding and management, they grow rapidly. Among the different breeds, they are considered the heaviest at any given age.

Spotted Poland China

The present breed of Spotted Poland China can be traced back to the spotted pigs whose ancestors were the original Poland China. The Spotted Poland China is of more recent origin than most of the other established American breeds of pig.

The standard colour of the Spotted Poland China is 50 per cent white and 50 per cent black with large, even spots. The Breed Association of the Spotted Poland China requires that the body must be at least 20 per cent but not more than 80 per cent white (Fig. 1.9).

Spotted Poland China pigs sometimes tend to become chubby with too much fat laid down on body and neck. For breeding purposes, strict selection should be conducted to produce meat-type pigs, especially when the Poland China is crossed with indigenous pigs in the Tropics.

A large number of farmers in the Tropics do not quite accept a spotted breed of pig and for breeding purposes they prefer a breed with a solid colour.

Tamworth

The Tamworth is said by some authorities to have originated from Ireland where the Irish called the pigs

fig. 1.10 Tamworth pig

Irish Grazers. Some of these pigs were brought to Stamford and Tamworth in England. Other authorities state that the foundation boar used for fixing the golden colour was a jungle pig imported from India around 1800.

The Tamworth may have been infused with Yorkshire and Berkshire blood. Sometime in 1882, the Tamworth was imported to the USA and Canada.

The Tamworth is long-bodied, long-necked and long-legged. It's most striking feature is its straight head (Fig. 1.10). The Tamworth colour varies from golden red to dark red.

The Tamworth is recognised as a bacon-type pig. However, crossbreds of the Tamworth and some other breeds produce excellent lean-type pork pigs. Tamworth pigs feed well on pasture, they are efficient converters of feed to liveweight, they are also hardy and less sensitive to sunburn than other temperate-type breeds. Many breeders prefer to use the Tamworth for breeding a lean-type pig because it is a thrifty, rugged and active breed.

Tables 1.1 and 1.2 present the production performance of some major breeds of temperate-type pigs reared in a tropical climate.

Table 1.1 Production performance of some major breeds of temperate type pigs in the Tropics. (*Sources:* **Mai, S. C.** (1966). Taiwan Sugar Corporation Hog Raising Program and its Progress. *Taiwan Sugar*, **13**, 74. **Argañosa, V. G. and Abilay, T. A.** (1963). The Berkshire swine in the College of Agriculture, UPLB. *Phil. Agriculturist*, **47**, 68–83. **Rigor, E. M., Peñalba, F. F., Beltran, E. D., Alcantara, P. F. and Abilay, T. A.** (1974). The further developments, multiplication and performance testing of an NSDB-UPLB developed strain of pigs and some selected purebreds. *Unpublished report*. **Trakulchang, S., Fajardo, R. C. and Rigor, E. M.** (1967). The performance of Philippine pigs and Philippine pig × Landrace crosses (F_1) II. Carcass evaluation, *Phil. Agriculturist*, **51**, 227–230.)

Breed	Country	Year	Average initial weight (kg)	Average final weight (kg)	Average daily gain (kg)	Feed efficiency (kg feed kg liveweight)	Backfat thickness (cm)
Native							
Taoyuan	Taiwan	1960–64	10·99	80·22	0·35	4·22	4·55
Ilocos	Philippines	1967	12·15	68·28	0·22	4·78	4·22
Berkshire	Taiwan	1960–64	14·70	80·23	0·41	3·53	4·06
	Philippines	1963	7·83	82·54	0·30	3·89	4·32
Yorkshire	Taiwan	1960–64	13·75	80·45	0·52	3·61	3·94
	Philippines	1973–74	15·00	87·00	0·63	3·74	2·87
Landrace	Taiwan	1960–64	16·14	80·43	0·57	3·51	3·43
	Philippines						
Duroc-Jersey	Taiwan	1960–64	15·59	80·46	0·61	3·49	3·78
	Philippines	1973–74	12·05	85·00	0·62	3·69	2·92
Hampshire	Taiwan						
	Philippines	1973–74	13·78	82·57	0·57	3·98	2·46

Table 1.2 Average breeding performance of different breeds of pig. (*Sources:* **Arganosa, V. G. and Guevarra, A. J.** (1974). Performance of swine in selected piggeries in the Philippines. VI. Reman Enterprises, Inc. (REI) Lipa City, 1974. *Better Poultry and Livestock,* **16**, 30–32. **Argañosa, V. G., Puyaoan, R. B. and de Ramos, M. B.** (1970). The performance of pigs born in different years and months of the year. *Phil. Jour. of Anim. Science,* **7**, 29–37. **Rigor, E. M., Peñalba, F. F., Beltran, E. D., Alcantara, P. F. and Abilay, T. A.** (1974). Further development, multiplication, and performance testing of NSDB-UPLB developed strain of pigs and some selected purebreds. University of the Philippines, College of Agriculture, Department of Animal Science. *Unpublished report.*)

Breed	Year	Number of pigs per litter			Weaning percentage based on:	
		at birth	born alive	at weaning	total number of pigs born	pigs born alive
Crossbred	1963–68	9·65	9·15	7·16	74·20	78·25
	1972–73	9·80		7·90	80·61	
	1974	9·30	8·80	7·80	83·87	88·64
Purebred						
Berkshire	1963	6·20		4·40	70·97	
Duroc Jersey	1971–74	9·30	9·04	6·64	71·40	73·45
Hampshire	1971–74	7·33	7·14	5·86	79·95	82·07
Yorkshire	1971–74	8·67	7·83	5·33	61·48	68·07
Native (Philippines)	1971–74	7·11	6·56	5·78	81·29	88·11

Tropical pigs

Central and South America

The different types of local coastal strains of pigs probably originated from pigs of Chinese ancestry introduced by the Spanish conquerors of South America as early as the fifteenth century. Alonso de Ojeda was the first man to introduce pigs to tropical America. In a letter to Diego Vasquez, the conqueror of Cuba, he mentioned that about 30 000 pigs had already populated the Island since the conquest. Pigs from Cuba soon reached Puerto Rico, Jamaica and Mexico. In about the same year, Captain Jaime de Pinzon brought to Colombia some of the strains which later became widely spread in the Antilles and along the Colombian coast.

In general the native pigs in tropical Central and South America are of the lard type. The **Pelon** is a small size Mexican breed with little hair; the **Zungo Costeño** and **Congo Santandereano** are common native breeds of Colombia and Venezuela; the **Pereira, Talú** and **Pirapitinga** are essentially hairless and small lard-type breeds indigenous to Brazil. The **Criollo** and **Nilo** are native strains which are generally found in other tropical countries of Latin America.

Southeast Asia

The sway-back **Hainan** pig of mainland China is noted for its mothering instinct (Fig. 1.11). It is said that mortality among baby pigs is very low because the sow lies down very carefully and slowly during nursing time.

The Laos native pigs, locally known as **Muban** are very similar to the pigs in the Philippines known as **Jalajala, Koronadal** and **Ilocos**. They are short and sway-back with a sagging belly, a long nose, a prominent snout and hairs that stand out on top of the shoulder and neck.

While they are mostly black in colour, some are spotted. They weigh 40–50 kg at 1 year of age. The number of pigs in their litter ranges from 3 to 7 at farrowing and 2 to 5 at 56-day weaning (cf. Fig. 1.12).

fig. 1.11 Hainan (Chinese) pig

fig. 1.12 Sway-back pig, Bali

fig. 1.13 Burmese pig

fig. 1.14 Akha pig, Thailand

There are 2 strains of pigs known to exist in Burma – the **Burmese** pig (Fig. 1.13) with its short nose and the **Chin dwarf** strain. Burmese pigs are generally black but some have spots. The Chin dwarf is short and weighs 30 kg at maturity.

In Thailand, the **Akha** pigs (Fig. 1.14) are important in market farming for subsistence.

fig. 1.15 Local pigs on their way to market in Cameroon

Africa

West African pigs are said to have originated from the Mediterranean region. A breed called **Bakosi** is found in Cameroun. They are generally black or spotted. Being good scavengers they easily survive by feeding on wild rootcrops (Fig. 1.15). In West Africa a breed of dwarf pigs is common (Fig. 1.16).

The bush pig (*Potamochoerus porcus*), wart hog (*Phacochoerus aethiopicus*) and the giant forest hog (*Hylochochoerus meinertzhageni*) are some of the wild members of the Suidae family found in Africa, but it is unlikely that any domestic breeds have been developed from them.

The **wart hog** has long legs, is 1·5 m in body length and has a tufted tail. The skin is slate or clay coloured and naked except for a few bristly hairs. It possesses a mane

fig. 1.16 West African dwarf pig

fig. 1.17 Wart hog

that is continued along the mid-line of the back. The muzzle is broad with upward-curving tusks (Fig. 1.17).

The **bush pig** is slightly smaller than the wart hog. The coat of short hair is reddish, but may be black in old males. The ears are long and pointed and end in white tufts. The tusks are small. It is found south of the Sahara.

The **giant forest hog** is the rarest and least known. It has been found in several parts of Kenya, such as the Aberdare mountains and Mount Kenya, in the mountain forests of north Tanzania, in Uganda and as far west as Liberia.

Further reading

Briggs, H. M. (1960). *Modern breeds of livestock.* McMillan: New York.

Eusebio, J. A. (1969). *The science and practice of swine production; with emphasis on Philippine conditions.* UPCA Textbook Board: Los Baños, Philippines.

National Spotted Swine Record (1965). *History of the spotted swine herd.* The National Spotted Swine Record: Bainbridge, Indiana.

Moore, F. F. and Moore, L. P. (1964). *The origin and history of Chester White Breed.* Chester White Swine Record Association: Rochester, Indiana.

Poland China Record Association (1963). *Using Poland China in commercial pork production.* Poland China Record Association: Gatesburg, Illinois.

United Duroc Swine Registry (1964). *Facts and hi-lites on Duroc.* United Duroc Swine Registry: Peoria, Illinois.

Williamson, G. and Payne, W. J. A. (1978). *An introduction to animal husbandry in the tropics.* 3rd ed. Longman: London.

Zeller, J. (1957). *The story of Landrace in America.* American Landrace Association: Springfield, Missouri.

2 Breeding for efficient production

Traits are inherited through genes. Simply inherited traits are governed by one or two pairs of genes. A pair of genes may be transmitted and expressed in one of two forms: dominant or recessive.

The dominant gene, regardless of whether it is paired with a recessive gene or with a similar dominant gene, will still express its effect. For example, white colour is dominant to black. Thus, when a Large White boar is mated to a black sow, all the offspring will be white. However, not all the white pigs in the litter carry the genes for pure white colour, and if a boar from this litter (first generation) is crossed with a gilt litter mate, the resulting offspring in the second generation will be on average 75 per cent white and 25 per cent black in colour. The black pigs, possessing recessive genes, will breed pure for black colour (Fig. 2.1).

Recessive genes do not show their effects in individual animals that also possess a dominant gene. They are transmitted from generation to generation without being noticed. The traits controlled by the recessive genes will only appear in the offspring if both a boar and a sow having the same recessive factor are mated. Recessive genes may produce either desirable or undesirable traits. Examples of the effects of undesirable recessive genes in pigs are scrotal hernia and inverted nipples. Thus, if these traits appear in the herd, both the boar and the sow as well as their litter should be culled as it is impossible to know which pigs possess the recessive genes (Fig. 2.2).

There are other traits, more complex in inheritance, that are of interest to pig breeders and producers. Such traits are rate of liveweight gain, feed efficiency, and carcass quality. These traits are controlled by many pairs of genes and the genetic process involved is very complex. It is known, however, that when an unrelated sow and boar are mated, the resulting offspring show better production performance than either parent. They are said to exhibit hybrid vigour. The less the parents are related the greater is this expression of hybrid vigour.

Breeding systems

Breeding systems are those methods for breeding animals, used by breeders. The terms outbreeding, crossbreeding and inbreeding are commonly used by breeders for the

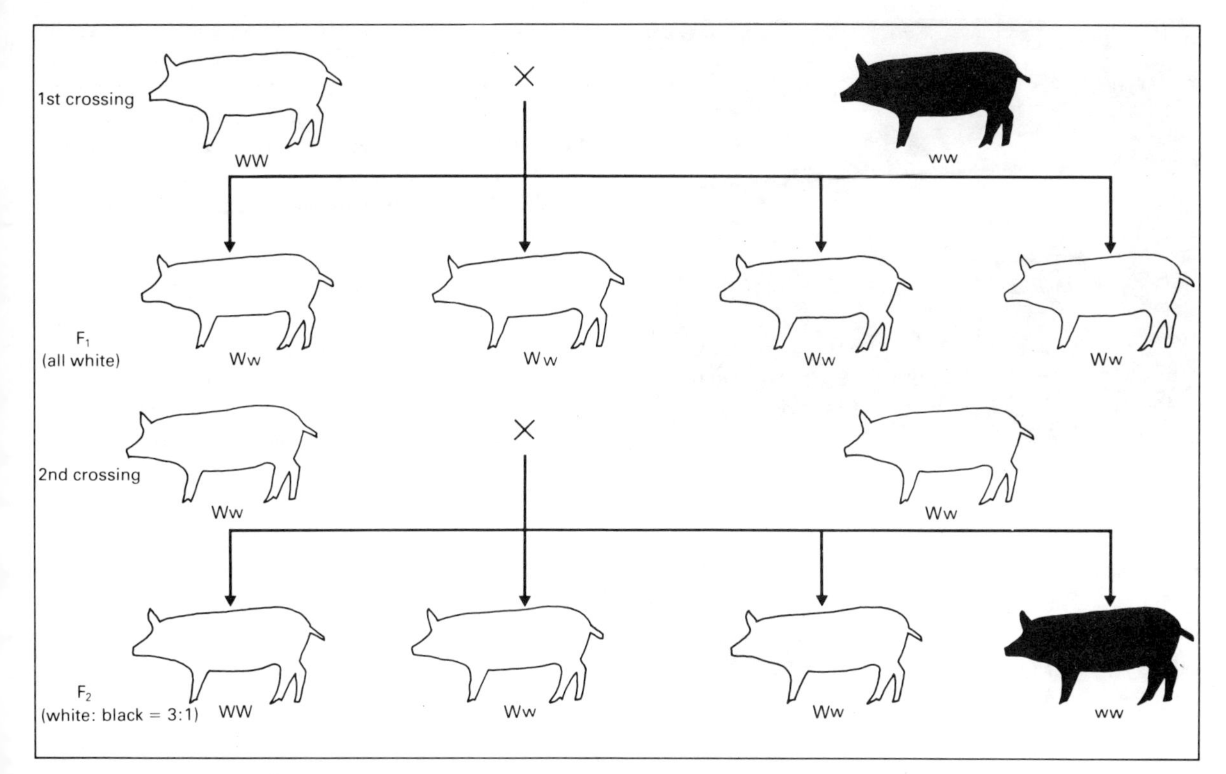

fig. 2.1 A dominant/recessive cross

system of mating their livestock. The system of mating is particularly important in pig improvement since use of the correct system can greatly increase production.

Outbreeding In outbreeding, the mating is brought about by pairing pigs that possess unlike or dissimilar ancestries or pedigrees. Outbreeding tends to produce litters with greater vigour and productivity.

Crossbreeding The expression of hybrid vigour of individual pigs that possess different groupings of the various pairs of unlike genes is increased to a maximum by crossbreeding. Some of the genes which influence vigour may exist in homozygous form or in pairs of the same genes. Crossbreeding results in heterozygous individuals in which most pairs of genes contain one of the dominant genes that influence vigour.

Inbreeding If the system of mating is brought about by pairing related individuals or those with a similar pedigree, it is called inbreeding. In this system, the similarity of the animals within the group is increased. Inbreeding usually decreases vigour because it brings together the recessive genes with undesirable effects in the

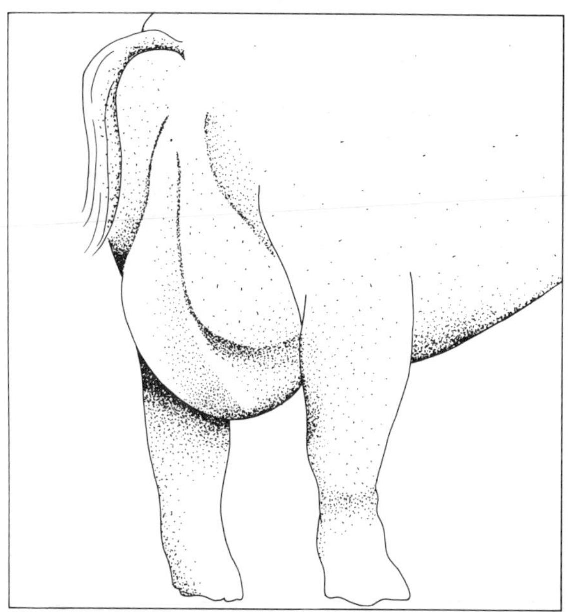

fig. 2.2 A scrotal hernia

resulting crossbreds. These homozygous recessive individuals are usually inferior. Inbreeding, however, is not always disadvantageous, even without selection in a herd, provided there are no undesirable recessive genes existing in the stock. If strict selection is practised, inbreeding may be useful for the purpose of eliminating defects. It brings out the desired character in a pure form and this character may then be retained.

Many improvements achieved in pig production during the past years have resulted from improved breeding and the use of productive breeds in upgrading, purebreeding and outcrossing, and to the practice of strict selection. This discussion on breeding will be limited to a consideration of crossbreeding for pork production, which is the most practical approach to increasing production in tropical countries (Fig. 2.3).

fig. 2.3 A Large White boar running with a small herd of Saddleback gilts. Their offspring will be used for pork production

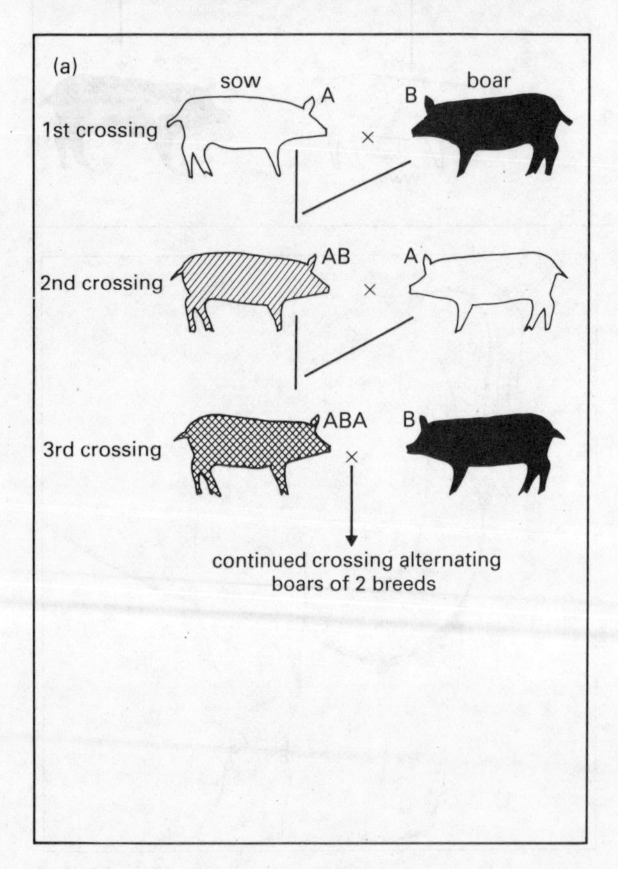

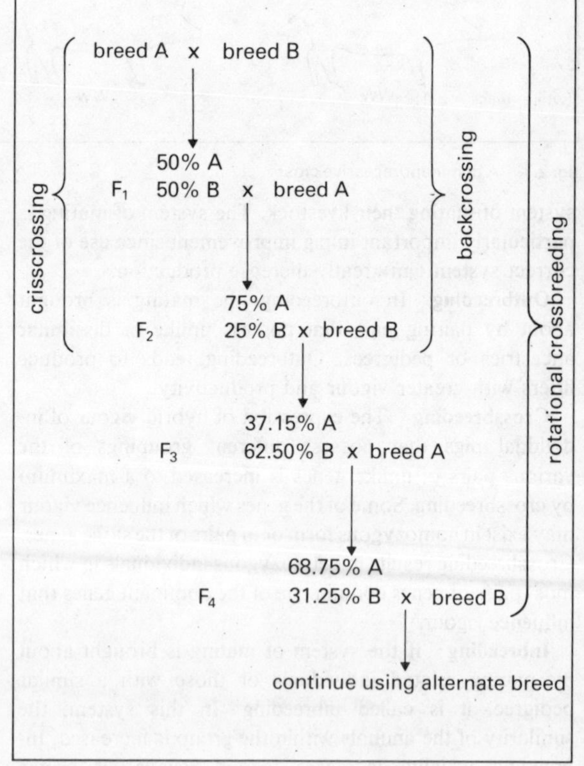

fig. 2.4(a) and (b) Crisscrossing of 2 breeds

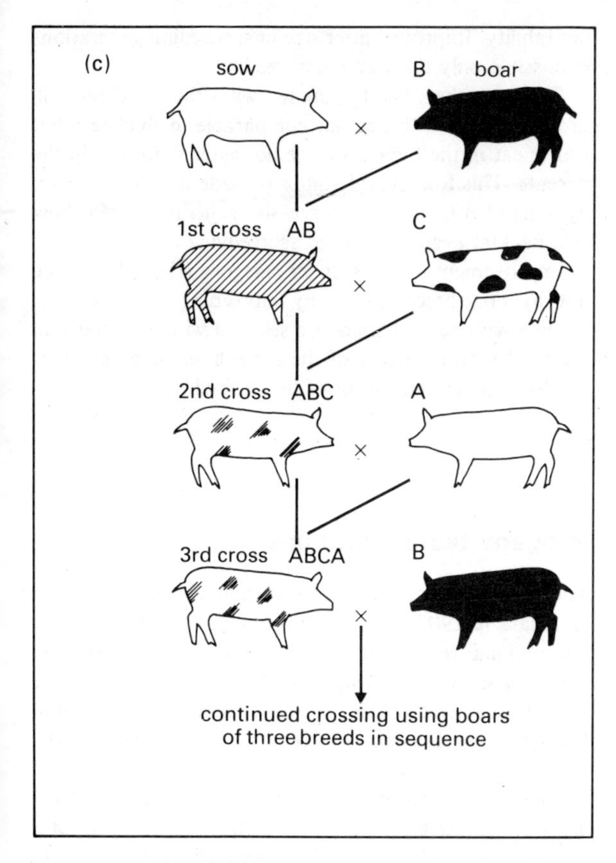

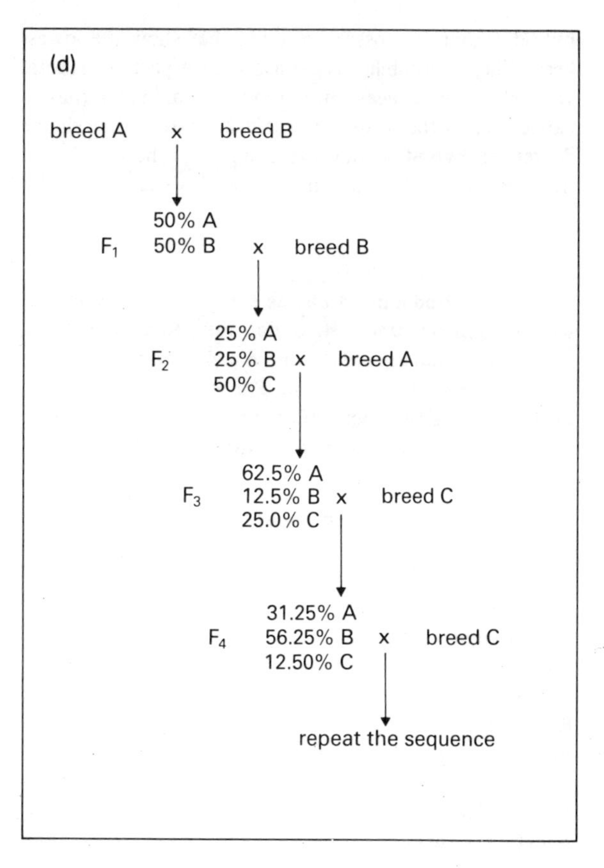

fig. 2.4(c) and (d) Triple crossing of 3 breeds

Crossbreeding for pork production

About 80 to 90 per cent of the pigs marketed by commercial and backyard farms in the Tropics are crossbreds. This large proportion indicates that both small- and large-scale pork producers have found crossbreeding to be an essential and profitable management practice.

Some examples of crossbreeding are crossing two different breeds, crisscrossing and triple crossing.

Crossing two different breeds involves the mating of purebred boars to purebred or high grade sows of another breed.

Crisscrossing, otherwise known as a two-breed rotation, means that boars of two different breeds are used in alternate generations. Crossbred sows are retained each generation and bred to boars of the same breed as the grandsire on the dam's side. Under this system, both sows and piglets are crossbred after the first crossing. Boars are purebred and come from a herd with proven high performance. Crossbred vigour usually results in an increase in litter size, livability and growth rate (Fig. 2.4(a) and (b)).

In **triple crossing** or a three-breed rotation, the first cross gilts are mated to a boar of the third breed. The process is continued by using a sire selected from each of the three breeds in rotation (Fig. 2.4(c) and (d)).

Multiple crossing is a system wherein boars of different breeds are used at specific intervals in order to prevent the production of pigs with excessive fat. By using boars of the bacon-type such as the Landrace crossbred strains, hybrid vigour and approved type are secured.

Hybrid vigour Traits that demonstrate the most

hybrid vigour in crosses are those that show the lowest heritability. Heritability is defined as that part of the total variability (differences in a group of animals) that is carried on to the offspring. Table 2.1 shows the relative degree of hybrid vigour expression and heritability of traits as indicated by the number of plus signs.

Table 2.1 Production traits as expressed in hybrid vigour. (*Sources:* **Baca, R. C. de, Rust, R. E. and Hazel, L. N.** (1961). Crossbreeding for the commercial pork producer. *Iowa State Univ. Expt. Sta. Pamphlet.* **Brunner, W.** (1962). The Ohio pork improvement agricultural extension service. *The Ohio State Univ. Bull.*)

Trait	Hybrid vigour	Heritability	Category
Litter size	+++	+	I
Litter weight	+++	+	
Pig livability	+++	+	
Rate of gain	++	++	II
Efficiency of gain	+	++	
Length of body		++++	III
Backfat thickness		+++	
Loin eye area		+++	
Lean cut		++	

Under category I are listed the traits that exhibit a high degree of hybrid vigour with a low heritability; category II includes traits with a moderate hybrid vigour and moderate heritability; while category III lists traits that exhibit no hybrid vigour but a high heritability. The implication is that only traits listed in categories I and II can be improved by crossbreeding.

For example, if a Berkshire boar coming from a litter of 7 is crossed with a Chester White sow from a litter of 6, it is possible to get a crossbreed litter of 9. This large litter size will probably not be passed on to the succeeding generations of the hybrid unless crossbreeding is repeated (as in crisscrossing). In general therefore, this implies that hybrid vigour in litter size is limited to only one generation following crossbreeding. This is due to its low heritability. Improved litter size in succeeding generations is possible only through crossbreeding.

On the other hand, parents with thin backfat will produce litters with backfat comparable in thickness but lean meat in the hybrid will be no better than that in the parents. This low hybrid vigour is made up for, however, by a high heritability; i.e. successive generations will show thin backfat even without crossbreeding.

Improvement of traits under category III are best done not by crossbreeding but by inbreeding. In this case, parents with desired traits are selected within a litter and are bred with members of the same litter or breed. This results in an enhancement of the desired trait.

Progeny testing stations

The first pig progeny testing system was developed in Denmark in 1907. This progeny testing system gained an international reputation for the development of excellent bacon-type pigs. During the 1920s, progeny testing spread to other countries in Scandinavia and continental Europe, New Zealand, Australia, Canada, USA and the UK.

Among the tropical countries of Southeast Asia, Taiwan was the first to establish a pig progeny testing station as early as 1960. The Taiwanese claim that in 1970, a total of 4·2 million head of pigs was slaughtered. This is about twice the number slaughtered in 1959, and the increase is considered partly to be the result of the improvement in their stock through progeny testing.

On account of the increasing demand for muscular meat-type pigs and the apparent decline in consumer acceptance of fat, the pig industry finds it necessary to produce meat-type pigs. Also, several countries in the Tropics producing pigs would like to sell their products in the world market.

The selection of meat-type breeding stock that will grow rapidly and efficiently is a major problem for commercial pork producers in the Tropics.

Progeny test for evaluating boars

The appraisal of a prospective breeding boar is an estimate of his breeding value. In making the appraisal or evaluation the following points should be considered:

1 the merit of the ancestors in his pedigree;
2 the merit of the boar himself;
3 the merit of collateral relatives, such as his full brothers or his half brothers and sisters; and
4 the merit of the individual's progeny after he is old enough to have been tested.

When this information is available, then the boar is selected or culled according to whether he can contribute towards the genetic improvement of the herd. The choice of herd boars is a most important decision because they will be responsible for a large proportion of the inheritance of the next generation of the herd. There is a greater opportunity to select for desirable characteristics in boars than in gilts. Therefore, it is desirable to use the best possible method of selection to increase the accuracy of evaluating the breeding value of prospective boars. The first decision must be made on individuality, pedigree and family information. However, these should be checked by adequate and accurate progeny tests.

Testing procedure for breeding boars.

Boars from the more productive sows are put on a feeding test after weaning (56-day old). The feeding test should last until the boars weigh 80 kg. Rate of liveweight gain during the test is measured for each individual, and feed efficiency is measured for 2 to 4 boar pigs coming from the same sire. The pen will usually only contain litter mate boars, but in some cases boars from two litters of comparable age and by the same sire are fed together. Efficiency of liveweight gain, expressed as the number of units of feed required per unit gain in weight is measured on a family group of full or half brothers. Boars under test are fed on a specific type of ration. They are weighed at intervals of 2 weeks until they attain 80 kg liveweight. The backfat thickness of each boar is measured by probing with a lean meter or by some other method. Each boar is also 'scored' for carcass length, meatiness and soundness of legs.

Reproduction and sterility in pigs

Farmers in many of the underdeveloped countries in the Tropics still find low reproductive efficiency in pigs and other farm animals. This setback in production can be attributed partly, if not totally to disease, the effects of the environment, poor nutrition and lack of technical expertise.

Pigs can reproduce at any time of the year and they are litter-producing animals. It has been estimated in some countries in Southeast Asia and South America that the first five pigs farrowed represent the total cost of producing a litter from fertilisation to weaning. This means that every effort should be made to strive for high fertility and the greatest number of pigs in farrow. The economics of pig production is based on the ability of the sow to reproduce and to raise her pigs efficiently at the estimated cost and in good time.

Oestrous period

A sow cannot conceive if mating does not take place during the appropriate period of the oestrous cycle. The length of the oestrous cycle ranges from 19 to 24 days, with an average of 21 days. During the time that the sow can conceive she is said to be 'in heat'; this is the **oestrous** period of the cycle.

The sow or gilt 'in heat' gives a characteristic grunt and/or shows restlessness. The external genitalia, the vulva, begins to swell. During this phase, the female will accept the male when he attempts to mate with her.

The gestation period of pigs ranges from 112 to 120 days, with an average of 114 days.

Causes of infertility in pigs

When gilts or sows are mated several times by normal fertile boars without becoming pregnant, the problem is likely to be either hormonal imbalance or abnormalities in the genital organs. In both cases the gilts or sows are considered infertile and should be disposed of as soon as possible. Although fertility in some tropical countries, especially in Southeast Asia and South America may appear to be within the normal range, strict culling should be practised to remove gilts or sows that do not conceive after the second mating. High ambient temperatures and a poor nutrition programme (e.g. high-energy feed intake during pregnancy) can lower the reproductive performance of breeding sows. It has been demonstrated in studies conducted in the Philippines that pregnant gilts require high-energy intake for the best reproductive performance, but after the first farrowing, energy intake should be minimised for good results.

Other causes of infertility in gilts or sows are due to disease. Brucellosis and leptospirosis are, respectively, the diseases most responsible for losses due to infertility and abortion. Although neither disease is common in tropical regions it is important to remain vigilant for any signs of disease, especially brucellosis, in the pig herd. Other bacterial infections that reduce reproductive efficiency are those of streptococcus, staphylococcus and salmonella organisms. Abortions caused by these organisms are usually sporadic in incidence.

Artificial insemination

The majority of pig raisers in underdeveloped tropical countries practise the backyard system of pig production in which one or two sows are kept. In the backyard system keeping a boar is not economic because of the cost of maintenance. Although a boar may be used to serve a number of sows in the community by natural mating, the disadvantage is that he may become a vector for the transmission of disease from one farm to another. Therefore, an efficient and progressive artificial insemination (AI) programme is of great value to small-scale pig farmers.

Some of the benefits of artificial insemination are listed.

1 It is a means of using outside boars in disease-free sow herds.
2 It permits breeding more sows by boars of proven merit.
3 It eliminates the problem of difference in size of males and females.

There are two major problems that prevent the widespread use of AI in pig production. These problems are associated with basic differences between the semen of bulls and boars. Bull semen can be diluted and will remain fertile for a week or more, and it can be frozen and kept indefinitely. Whereas there is a rapid decline in the viability of boar semen after 1 day of storage, and in most cases it loses all fertilising capacity after storage for more than 2 days. While bull semen may be greatly extended by proper dilution, with an average ejaculate providing enough diluted semen to inseminate as many as 500 cows, an average boar ejaculate can only be diluted to inseminate 8 or 9 sows.

Collecting the semen

To be successful in semen collection, it is necessary to simulate for the boar the natural conditions found in the female reproductive organ. The temperature and the pressure found in the female reproductive organ must be applied to the penis while the semen is being collected (Fig. 2.5(a) and (b)).

There are two popular methods used in semen collection:

The artificial vagina There are several models, among which the Bovine or Russian, the Japanese and the McKenzie types are commonly used. These artificial vaginas are made of rubber tubing reinforced at the mouth with a collecting bottle at the end. In order to collect the semen the boar must be trained until experienced in the operator's use of the artificial vagina.

The electro-ejaculator This method makes use of an electric shock to stimulate ejaculation. The boar does not need training.

In general, the volume of semen per ejaculate increases with age but decreases as collection becomes more frequent. The average volume of ejaculate is 200 ml. The sperm concentration is usually 240 million per ml. Eight per cent of the semen consists of a liquid and the rest consists of the gelatinous portion.

Semen processing

Boar semen must be protected from rapid changes in temperature immediately after collection. It should be cooled slowly to the temperature at which it is to be stored. If only a limited number of sows are to be inseminated, there may be enough semen to provide 50 ml of undiluted semen per sow.

On the other hand, dilution is necessary if one ejaculate is to be used to provide semen for a number of sows; especially if the insemination service is extended to the rural areas. About 1 part of semen to one part of diluent is the normal dilution. If it is diluted, 200 ml of semen in one ejaculate can be extended to 8 sows.

Two diluents or extenders have been tried and tested in the Philippines.

Illini variable temperature (IVT) The preparation of this diluent essentially involves the use of the IVT buffer without egg yolk. In order to prepare it the following chemical ingredients should be weighed and placed in a container with not more than 500 ml of distilled water.

20 g of sodium citrate dehydrate
2·1 g of sodium hydrogencarbonate
0·4 g of potassium chloride
3·0 g of glucose
3·0 g of sulphanilamide

Heat gently and stir to dissolve the chemicals. Cool and then make up the volume to 1 litre with more distilled water. Transfer the prepared solution to a coloured bottle or container, to protect from light, and store this in a cool dry place until it is to be used.

Cornell University extender (CUE) To prepare a litre of this extender, weigh:

14·5 g of sodium citrate dehydrate
2·1 g of sodium hydrogencarbonate
0·4 g of potassium chloride
3·0 g of glucose
3·0 g of sulphanilamide
9·33 g of glycine
0·87 g of citric acid crystals

Then follow the same procedure used in the preparation of the IVT diluent.

Dilution of boar semen

Pour the desired volume of the prepared diluent into a container; preferably a calibrated and stoppered cylinder. Bubble carbon dioxide gas through the liquid for 10 minutes and add antibiotics. Pour approximately the same amount of newly collected boar semen into the solution (buffer and antibiotics). Store this diluted semen at a temperature of 5°C. It is essential to add antibiotics before adding IVT or CUE diluent to semen. Penicillin and streptomycin are the two most common antibiotics used for preserving semen. The recommended level for penicillin is 500 international units (IU) per ml of diluted semen and that for streptomycin is 500 mg per ml of the diluent.

Insemination

The mechanics of insemination are probably the simplest of all the procedures involved in AI. A disposable plastic pipette, as used for cattle, but with a tip modified by a short bend which aids in the penetration of the cervix, is used to connect the pipette to a 50 ml syringe.

Most sows in heat will stand readily during the process of insemination with little need to restrain them (Figs 2.6

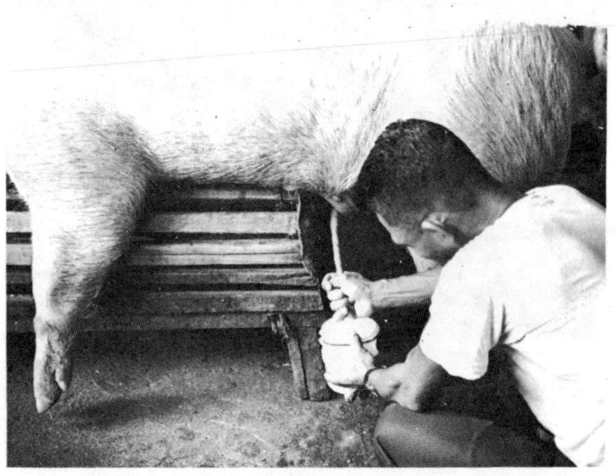

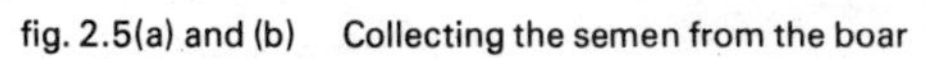

fig. 2.5(a) and (b) Collecting the semen from the boar

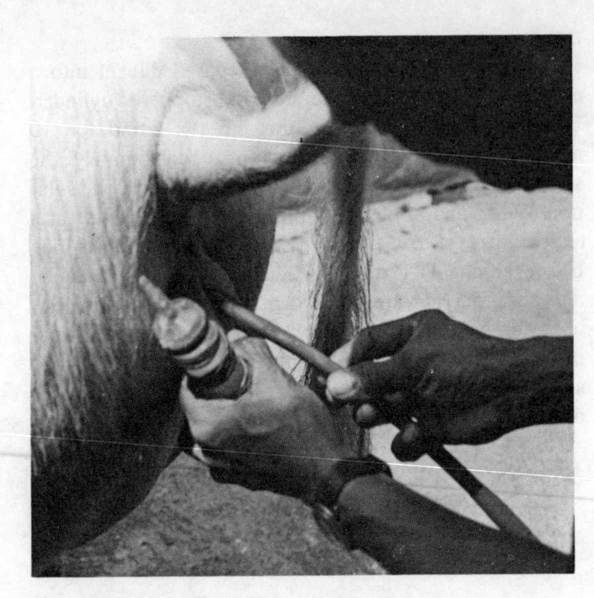

fig. 2.6 Insertion of catheter in vulva of sow

fig. 2.7 Insemination of sow

and 2.7). Perhaps the most appropriate time to inseminate a sow in heat is late during the 1st day or early during the 2nd day from the onset of oestrous. The chart in Table 2.2 shows the appropriate insemination period during the heat period of the sow or gilt.

Table 2.2 Breeding chart for sows or gilts. (*Source:* **Clamohoy, L., Supnet, M. G. and Palad, O. A.** (1967). Artificial insemination in swine. *Tech. Bull.,* **16.** UPCA: Los Baños, Philippines.)

Pre-heat	Standing heat	Diminishing heat
Gilts and sows show external signs of heat	1st day/2nd day (from onset of sexual receptivity)	3rd day (from onset of sexual receptivity)
There is a refusal to be mounted or tested for sexual receptivity	Receptive during this period	There is a refusal to be mounted or tested for sexual receptivity
Too early: poor result	Best time: good result	Too late: poor result

Further reading

Basco, C. (1953). *A study of the breeding habits of Berkjala swine.* Philippine Agriculturist, **35,** 408–414.

Turman, E. J. (1966). *Artificial insemination and oestrous synchronization in swine.* Oklahoma Swine Breeders Association: Oklahoma City.

Warnick, A. C. (1965). *Swine production in Florida.* State of Florida Dept. of Agric. Bull No. 21, Gainesville: Florida.

Part 2 Pig biology

3 Reproductive and adaptive physiology

Female reproductive system

The reproductive system of the female pig is composed of two ovaries, in which each egg cell or ovum, is produced, and a duct system, which receives the ova. Fig. 3.1 illustrates the different parts of the reproductive system of the female pig.

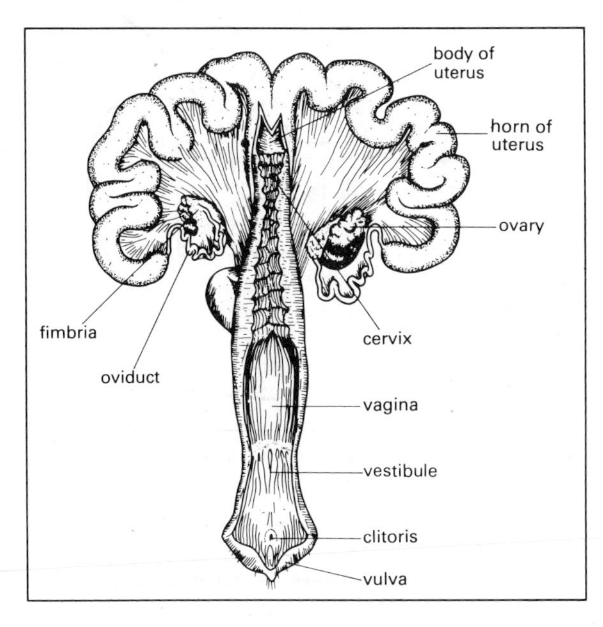

fig. 3.1 Reproductive organs of the sow

Ovaries

The **ova** originate from the epithelial covering of the ovary, the **germinal epithelium**. Normally, all the ova that a mature female pig will shed have been present in the ovary since birth.

Each **ovum** is enclosed by a **follicle** which contributes to the nutrition of the ovum. At birth, the ovum occupies most of the space in the follicle, and this is called a primary follicle. The follicle grows more quickly than the ovum, and the cavity within the follicle becomes enlarged and filled with liquid. As a result the follicle bulges at the surface of the ovary and becomes what is known as a **Graafian follicle**.

The **membrana granulosa** is the group of cells lining the cavity occupied by the ovum in the follicle. As the time for ovulation draws near, the outer wall of the cavity becomes thinner, and the follicle finally ruptures to release both the liquid and the ovum. When this takes place, the membrana granulosa is thrown into folds, and the follicular cells as well as those of the surrounding connective tissue become pale yellow staining cells. At this stage, the structure is called the **corpus luteum**, or yellow body. If the ovum is fertilised the corpus luteum persists and plays an important role in the ensuing pregnancy. If the ovum is not fertilised the corpus luteum is reabsorbed into the tissue of the ovary.

The duct system

The duct system of the pig is composed of the oviducts or fallopian tubes, the uterus, the cervix, the vagina and the external genitalia.

The **oviducts** serve as passageways for the ova from each ovary to the uterus. After the egg is expelled from the ovary, it is drawn into the oviduct through the **fimbria**. This is funnel-shaped and has a larger diameter than the oviduct. It is lined with small hairlike projections called **cilia**, which produce wavelike movements that conduct the ovum into the oviduct. There are also cilia inside the oviduct which move the egg into the uterus. In addition, the oviduct itself moves quite considerably and this is one reason why there are cases where eggs from one ovary wander into the opposite oviduct.

From the oviducts, the egg goes into the **uterus**. The uterus is divided into a body and two horns. The body is very short, about 5 cm long, while the horns are extremely long (1·2 to 1·5 m) and folded. The uterus is the site of implantation for the fertilised ovum and the walls are specially structured for bearing the foetus. This portion is thick and muscular and is the site of the contractions during parturition or birth. The walls are lined with mucus and many blood vessels, which serve to circulate blood between the growing foetus and the mother. The neck of the uterus, the cervix, is remarkably long (10 cm); it is lined with mucus like the uterine body and is surrounded by muscle tissue.

The **vagina** is a muscular tube about 10 to 12 cm long which acts as the receptacle for the male organ, the **penis**, during mating. The vagina also serves as a passageway for the foetus during birth, the elasticity of its wall allowing the necessary expansion. The foetus is moved down the vagina by means of contractions in the abdominal wall.

Posterior to the vagina is the **vulva**, the term used for the female external genitalia. The **urethra** is found in the vulva region leading from the urinary bladder. Unlike in the male reproductive system, however, there is no interconnection between the urinary system and the reproductive tract. In front of the opening of the urethra is a small erectile tissue, the **clitoris**, which is a major area of stimulation during mating. Its analogue in the male reproductive system is the penis.

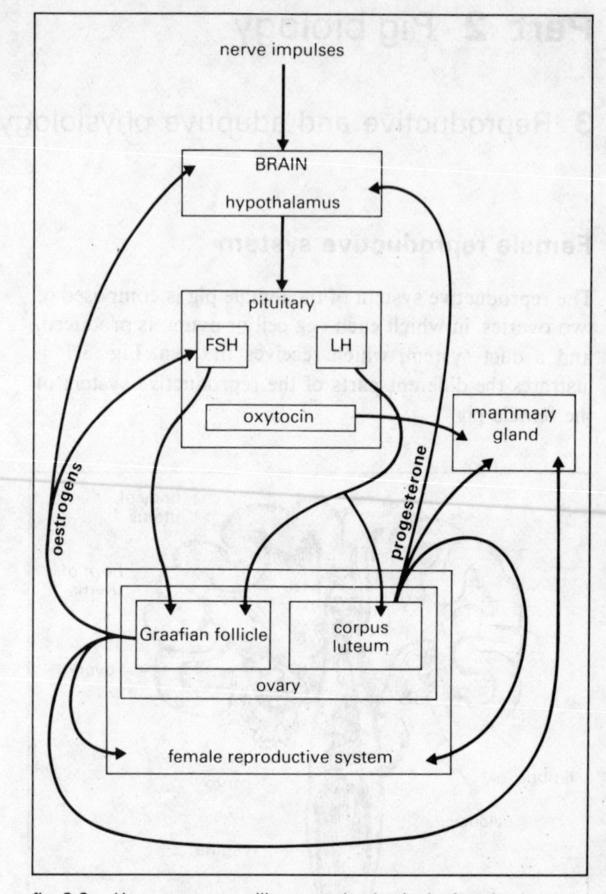

fig. 3.2 Hormones controlling reproduction in the female pig

The oestrous cycle

Like other farm animals with well-developed reproductive systems, the female pig is known to undergo a cyclic process which makes her periodically receptive to mating. This rhythmic variation involves a complex interaction between the different endocrine glands, which secrete hormones, and many of the parts of the reproductive system. The cycle is known as the **oestrous cycle**.

The first oestrous cycle is initiated with the onset of puberty. An endocrine gland at the lower part of the brain, the pituitary, is stimulated by the hypothalamus to release follicle-stimulating hormone (FSH) and luteinising hormone (LH). These, in turn, cause the ovaries, themselves endocrine glands, to mature and to secrete the female sex hormones, namely progesterone and oestrogen. The secretion of these hormones then initiates the growth of the secondary sex characters such as development of the mammary glands. These marked changes in hormonal activity and physical appearance of the maturing gilt occur at about 6 months of age and are part of a gradual process in growth development which is called **puberty**.

The oestrous cycle involves a synchronisation of the sex hormones by the female pig (Fig. 3.2). The sow can only be fertilised at the stage of oestrous cycle which is characterised by an increase in the thickness of the uterine wall and by an increase in the sow's sexual urge; the sow is then said to be 'in heat'.

Stages in the oestrous cycle.

1 *Proestrum*—the anterior lobe of the pituitary gland secretes FSH, which causes the growth of the follicles. As the follicles grow, they secrete increasing amounts of oestrogen, such that the overall amount in the blood reaches a level sufficient to inhibit further secretion of FSH. However, the oestrogen also causes secretion of LH by the pituitary.

2 *Oestrus*—when the amount of LH has reached its critical level, the follicle, now the Graafian follicle, ruptures and releases the ovum.
3 *Metoestrum*—the rupture of the Graafian follicle leaves a structure called the corpus luteum. This is a temporary endocrine gland and it continues to secrete oestrogen, although in a smaller amount, and the sex hormone, progesterone. At this stage, the vagina loses most of the growth it has made and the walls become thinner. The wall of the uterus also loses much of its mucus layer and some of the blood vessels are broken.
4 *Dioestrum*—progesterone is the hormone of pregnancy and prepares the uterus to receive the fertilised ovum or embryo. It has been shown that attachment of the embryo will not take place if progesterone is absent. Under its influence, the glands and blood vessels in the uterine walls mature. At this stage, the uterus is ready for implantation by the fertilised ovum or embryo and if this occurs, dioestrum remains throughout pregnancy and the corpus luteum continues to secrete progesterone.

If fertilisation does not occur, the pituitary begins to secrete less and less of the LH until the corpus luteum, which is maintained by LH, degenerates. Since the corpus luteum secretes progesterone, the progesterone level in the blood also falls and the uterine walls lose their thickness. The extra tissue and cellular material accumulated in the uterine wall is reabsorbed into the bloodstream; this is in contrast to the human cycle where the material would pass out of the body through the vulva. With the degeneration of the corpus luteum and the absence of progesterone in the blood, the pituitary then begins another oestrous cycle.

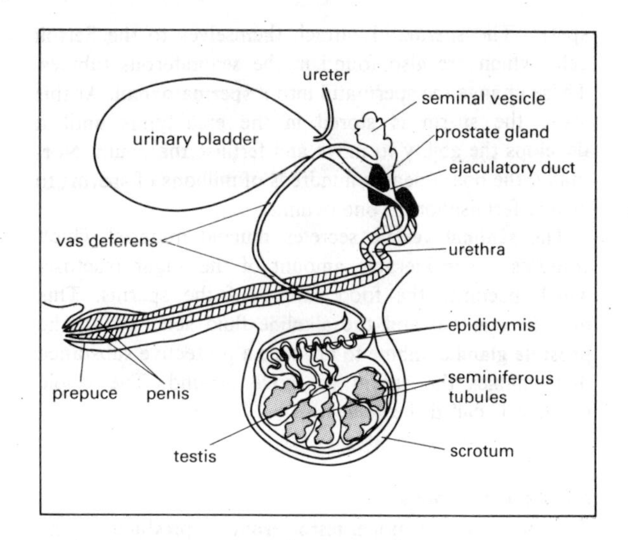

fig. 3.3 Reproductive system of the boar

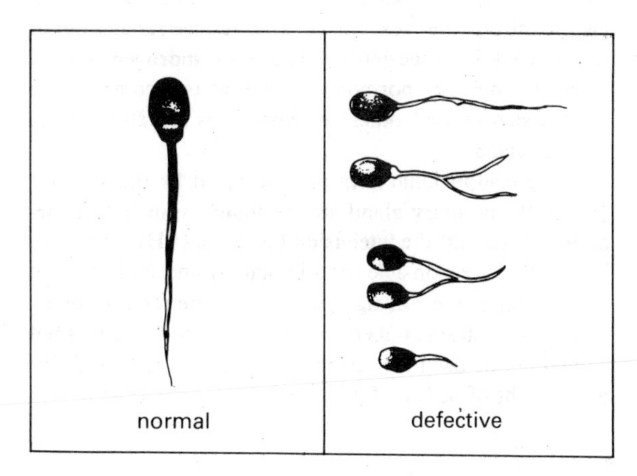

fig. 3.4 Normal and defective sperm

Male reproductive system

There are three reproductive processes in the boar: sperm formation or spermatogenesis; introduction of sperm into the sow's vagina during mating; and the hormonal control of sexual function. Fig. 3.3 illustrates the different parts of the boar's reproductive system.

Testes

Each testis is composed of a large number of seminiferous tubules where spermatogenesis takes place (Fig. 3.4). Spermatogenesis may continue normally throughout the entire life of the boar but after middle age, the tubules may degenerate or become atrophied. As the boar approaches senility, normal spermatogenetic activity declines.

The sperm

The first stage of sperm development is a **spermatid**; later in its life the spermatid grows a whip-like flagellum. This later becomes its tail when it is transformed into a mature sperm. The spermatids attach themselves to the Sertoli cells which are also found in the seminiferous tubules. These change the spermatid into a **spermatozoan**. At this stage the sperm is stored in the epididymis until it develops the ability to move and fertilise the ovum. Normally, the boar releases hundreds of millions of sperms to ensure fertilisation of one ovum.

The **seminal vesicle** secretes mucoid material which contains a considerable amount of the sugar fructose, which becomes the food supply of the sperms. This mucoid material and the alkaline fluid secreted by the **prostate gland** combine to produce a protective substance that enables the sperm to move around. The whole medium is called the **semen**.

Male sex hormones

The major sex hormone, **testosterone**, is produced in the Leydig or interstitial cells. This hormone is responsible for the growth and maintenance of the boar's sexual characteristics such as a deeper voice and rough skin; it also controls the sex drive. The female analogue to testosterone is progesterone. The term **androgen** is used when the male sex hormone referred to is a combination of testosterone and other sex hormones secreted by the adrenal glands.

The gonadotrophic hormones secreted by the anterior lobe of the pituitary gland are the **follicle stimulating hormone (FSH)** and the **luteinising hormone (LH)** (Fig. 3.5). The latter is responsible for stimulation and FSH for the maturation of the Leydig cells that secrete testosterone.

The boar attains puberty at the age of 4–6 months but should not be used for breeding until about 12 months or at a weight of at least 90 kg.

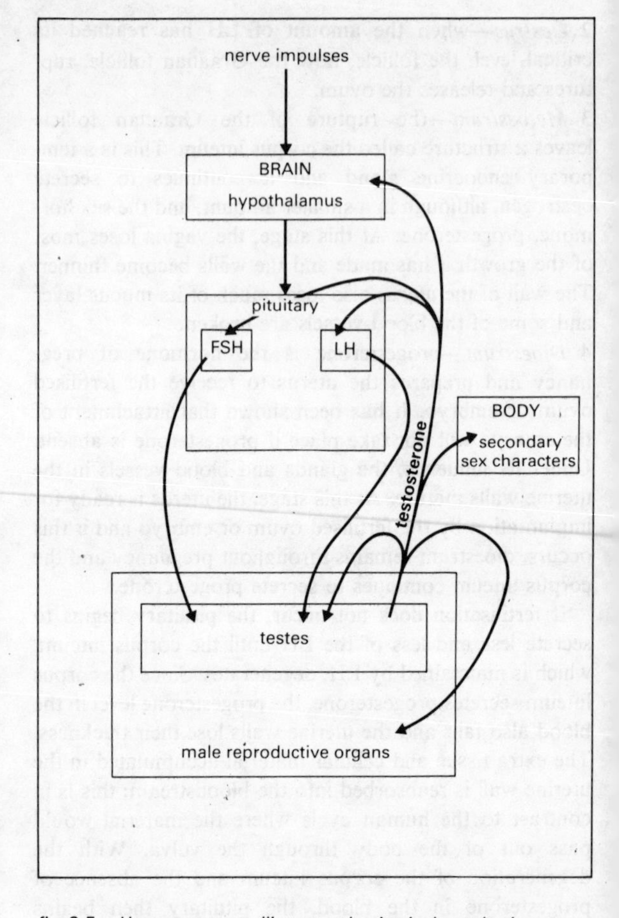

fig. 3.5 Hormones controlling reproduction in the male pig

Efficiency of reproduction in the sow

Age The age of the pig affects the rate of ovulation or the number of eggs released per oestrous cycle. The rates from the 3rd to 4th heats after puberty are higher than those in the first 2 heats. Studies have also shown that sows farrowing for the first time late in life produce 2 to 3 piglets less than those farrowing at the normal age (10–12 months). On the other hand, it has been reported that sows in their 5th or 6th pregnancies produced 1·90 and 1·92 more pigs than those pregnant for the second time.

Nutrition The number of ova released by the sow per oestrous cycle also depends on the nutritional status of the pig. This is shown by the high tendency for the fertilised egg to die (referred to as embryonic mortality) during pregnancy because the sow is either very old or overfed.

Hormones The reproductive efficiency of both boar and sow is also influenced by the sex hormones and the rate at which the hormones are secreted. However, this can be controlled by the proper genetic selection of parent stock.

Adaptive physiology

The complex environment of animals can be considered under two headings: the thermal and the non-thermal environments. The thermal part includes the environmental temperature, humidity and air movement or wind velocity, while the non-thermal part includes light, sound, odour, food and density of animal population in a given area. It must be remembered, however, that none of these factors individually affects the animal's normal physiological activities. The net change in the animal's activity is brought about by the interaction of a number of different factors.

The pig cannot sweat in response to high environmental temperature and therefore loses less water from its skin than most other domestic animals. This results from the fact that the skin of the pig is not richly supplied with blood and possesses no exocrine sweat glands. The pig's sweat glands are **apocrine** in type.

Physiological and behavioural responses may differ among animals even in similar environments. For instance, in the case of domesticated pigs, the young are most susceptible to a cold environment while mature boars or sows exhibit poor adaptability to a humid tropical climate. Among wild pigs, the mature animals cool themselves off by wallowing in mud while the piglets gather dried grass and leaves to make themselves warm and dry in cold weather.

Development of the thermoregulatory system of pigs

Baby pig phase Young pigs possess very little thermal insulation outside the skin surface and, therefore, any change in the environmental temperature can cause a marked effect on the skin temperature, especially during the first week after birth. Pigs are born in a litter of 6–12 and there is competition among the piglets for their mother's milk (Fig. 3.6).

The underweight and usually weaker piglets cannot compete with the stronger pigs in a litter for the most productive teats of the sow. As a consequence, the smallest pigs have low levels of metabolic capacity and poor resistance to cold. Thus, the weakest pigs can develop a condition called **hypoglycaemia**, which is characterised by a low level of blood sugar. The rate of

fig. 3.6 A dwarf West African sow feeding a litter

development of hypoglycaemia depends on the environmental temperature. At 15°C the animal is likely to die in 28 hours, while at an ambient temperature of 31°C the pig may only start to show a decrease in the blood sugar content after 8 hours. This trend suggests that warmer environments conserve the body carbohydrate reserves by reducing the loss of body heat of the pig. Under these circumstances good husbandry dictates that baby pigs should be kept warm even in a tropical environment.

Growing phase This phase begins at the time of weaning and lasts until about 6 months of age. During this phase, the growth and development of the pig is rapid, and there is vigorous muscular activity. Good husbandry in the growing phase is especially important because improper feeding and management practices can cause unwanted increases in the content of body fat, especially towards the latter part of the growing stage.

Because pigs do not sweat effectively under hot conditions their adaptation to a warm climate is quite difficult. One way in which pigs adapt to a warmer environment is the reduction of food intake. It has been reported that the metabolic activity of the pig may decrease during prolonged hot periods. This may be a physiological adaptation or it may be only a response to reduced feed intake.

Pigs can also adapt to cold climates despite their thin layer of hair coat. This is shown by the fact that pigs can live outdoors in the temperate zone with only dry bedding and the protection of a roof. The explanation is that pigs possess the ability to gain additional body fat, even if the diet provided for them in a cold climate does not normally allow for fat storage. Also, as an additional adaptation, young pigs kept at 10°C are reportedly able to grow more hair than those kept at a warmer temperature.

Mature and reproductive phase The 3rd stage of growth is marked by sexual maturity and reproductive activity. Compared to the first 2 life stages, the pig has fully developed its thermal insulation. With increased fat deposition, the pig can live in colder environments with little discomfort. However, the pig is less adaptive to hotter environments, especially those whose mean ambient temperatures are above 30°C. Such environments can, however, be ameliorated by the provision of a fine water spray or a wallow. If these are provided the pig can adapt and will exhibit a relatively improved growth performance and improved reproductive efficiency (cf. p. 80).

Environmental requirements for good husbandry

During the suckling stage of baby pigs, it is practical to let the pigs loose in an open pasture area where they can obtain some form of supplementary iron from the soil and the vegetation. Baby pigs are likely to suffer from lack of iron because the iron content in sow's milk is relatively low. However, care must be taken to make certain that the piglets do not become infested with parasites such as kidney and intestinal worms.

Under the closed-confinement system of raising pigs, the house must be well-ventilated, especially under humid tropical conditions. Where ventilation is poor there is often a high mortality rate among young pigs, because poor ventilation provides a very suitable environment for the development of various respiratory infections.

The air temperature inside the house can be partially controlled by the type of roofing materials and insulation used. In the Tropics, a roof made from palm-leaves or a grass such as *Imperata cylindrica* is desirable and produces a cooler air temperature inside the house. Aluminium is also a desirable roofing material but it is quite expensive. If the humidity inside the house exceeds that outside, this is probably due to the evaporative loss from the animals or from water in the internal drainage canals, and poor ventilation.

Further reading

Dukes, H. H. (1970). *Dukes' physiology of domestic animals*, 8th ed. Comstock Pub. Assoc.: Ithaca, USA.

Hafez, E. S. E., Sumption, L. J. and Jakway, J. S. (1962). The behaviour of swine. *The behaviour of domestic animals*, pp. 334–369. ed. E. S. E. Hafez. Bailliere, Tindall and Cox: London.

Hammond, J. (1954). *Progress in physiology of farm animals*. Butterworths: London.

Mount, L. E. (1968). *The climatic physiology of the pig*, pp. 1–291. Williams and Wilkins: Baltimore.

Scheer, B. T. (1963). *Animal physiology*. Wiley: New York.

4 Nutrition and feeding

The nutritional aspects of pig management involve a consideration of the digestive system of the pig and the factors affecting digestion, feed composition, pre-treatment of feeds and nutritional requirements during the different stages or phases of life, the formulation and balancing of nutrients in pig rations and the feeding management essential for efficient pork production.

Recent information on the nutritional requirements of pigs raised under tropical humid conditions was reported in the Philippines using locally available feedstuffs and management practices (Tables 4.1(a) and (b)). However, for purposes of comparison, the complete nutrient requirements for swine managed in non-tropical conditions are also presented in Table 4.1(c).

Table 4.1(a) Nutrient requirements of swine, amount per kilogram diet or per cent. *Source:* **Eusebio, J. A., Liebholz, J., Mendoza, R. B., Eusebio, E. C., Caparas, L., Supnet, M. G., Gatmaitan, O. M. and Momoñgan, V. G.** (1977). Nutrient requirements of swine under tropical humid conditions. *National Science Development Board – National Research Council of the Philippines Tech. Bull.* **2** (2) pp. 1–29: Los Baños, Philippines.

Liveweight (kg)	5–10	10–30	30–60	60–90
Average daily gain (kg)	0·26	0·32	0·42	0·60
Average daily feed* (air dry), g	450·00	745·00	1 730·00	2 430·00
Energy and protein				
Digestible energy (kcal)	3 500·00	3 500·00	3 500·00	2 900·00
Crude protein (%)	22·00	17·00	17·00	13·00
Inorganic nutrients (%)				
Calcium		0·90	0·60	0·60
Phosphorus		0·60	0·40	0·40
Sodium chloride		0·60	0·80	0·80
Sodium		0·23	0·31	0·31
Chlorine		0·36	0·48	0·48
Potassium		0·25	0·30	0·30
Vitamins				
Vitamin A (IU)		4 400·00	3 500·00	2 600·00
D (IU)		440·00	440·00	400·00
B_2 (mg)		2·00	2·00	2·00
B_{12} (μg)		22·00	22·00	11·00
Pantothenic acid (mg)		13·00	13·00	11·00
Antibiotics (g)		2·00–2·50	2·00–2·50	

* *Ad libitum* feeding from 5–30 kg liveweight and a scheduled feeding thereafter until market weight of 90 kg.

Table 4.1(b) **Nutrient requirements of breeding swine under tropical humid conditions.** (*Source:* as Table 4.1(a).)

	Pre-gestating		Breeding/gestating		Lactating		Boar	
	Sow	Gilt	Sow	Gilt	Sow	Gilt	Young	Mature
Liveweight (kg)	130–200	100–110	150–250	110–135	150–200	135–200	100–150	150–250
Daily feed intake (g)	3 000	3 000	2 800	2 500	3 000	3 000	2 800	2 800
Nutrient requirements, per cent of the diet or per kilogram								
Energy/protein								
Digestible energy (kcal)	3 016	3 016	2 325	3 400	3 016	3 370	3 035	2 325
Crude protein (%)	15	15	15	11	15	16	15	14
Vitamin A (IU) per kilogram				4 100		4 100		
D (IU) per kilogram				138		138		
Daily nutrient requirements								
Energy/protein								
Digestible energy (kcal)	9 100	9 100	6 500	8 500	8 100	10 100	8 500	6 500
Crude protein (g)	420	420	420	270	420	480	420	390
Vitamin A (IU)				10 660		14 350		
D (IU)				370		485		

Digestive system

In many ways the physiology of digestion and absorption in the pig is similar to that of man. Like man, the pig is **omnivorous**, consuming food of both plant and animal origin.

The alimentary tract or canal of the pig extends from the lips to the anus. Major parts of the alimentary canal are the mouth, pharynx, oesophagus, stomach, small intestine and large intestine (Fig. 4.1). Its function is to provide the necessary environment for the digestion and absorption of the food. Digestion involves the breaking down of complex foodstuffs into simpler compounds so

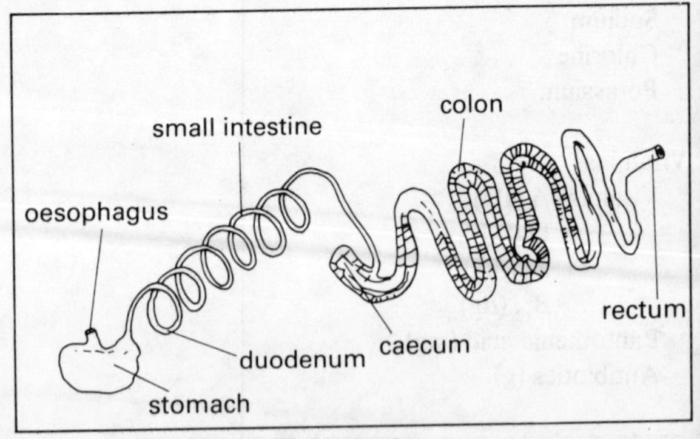

fig. 4.1 The alimentary canal of the pig

Table 4.1(c) Nutrient requirements of breeding and growing/finishing pigs. (*Source:* US National Research Council – National Academy of Science. (1968). *Nutrient requirements of swine,* No. 2, 6th ed. Publ. 1959. National Research Council: Washington.)

Nutrient	Growing		Finishing meat or bacon type	Bred gilt or sow (liveweight range,	Lactating gilt or sow (liveweight range	Boar (young adult) (liveweight range
	10–20 kg	20–50 kg	60–100 kg	100–160 kg)	140–200 kg)	140–200 kg)
Crude protein, %	18·00	16·00	14·00	14·00	15·00	14·00
Digestible energy, kcal	3 500·00	3 500·00	3 100·00	3 300·00	3 300·00	3 300·00
Minerals						
Calcium, %	0·65	0·65	0·50	0·75	0·60	0·75
Phosphorus, %	0·50	0·50	0·40	0·50	0·40	0·50
Sodium chloride, %	0·25	0·50	0·50	0·50	0·50	0·50
Vitamins						
Vitamin A, IU	1 750·00	1 300·00	1 300·00	4 100·00	3 300·00	4 100·00
β-carotene, mg	3·50	2·60	2·60	8·20	6·60	8·20
Vitamin D, IU	200·00	200·00	125·00	275·00	220·00	275·00
Thiamine, mg	1·10	1·10	1·10	1·40	1·10	1·40
Riboflavin, mg	3·00	2·60	2·20	4·10	3·30	4·10
Nicotinamide, mg	18·00	14·00	10·00	22·00	17·60	22·00
Pantothenic acid, mg	11·00	11·00	11·00	16·50	13·20	16·50
Pyridoxine, mg	1·50	1·10				
Vitamin B_{12}, mg	15·00	11·00	11·00	13·80	11·00	13·80
Choline, mg	900·00					
Aminoacids %						
Arginine		0·20				
Histidine	0·27	0·18		0·20		
Isoleucine	0·76	0·50	0·35	0·43		
Leucine	0·96	0·60		0·66		
Lysine	1·20	0·70	0·50	0·49		
Methionine	0·80	0·50		0·35		
Phenylalanine		0·50		0·52		
Threonine	0·70	0·45		0·42		
Tryptophan	0·18	0·13	0·09	0·08		
Valine	0·65	0·50		0·46		

that they can be absorbed through the wall of the intestine into the bloodstream or lymphatic system and carried to the parts of the body where they are needed. Enzymes are chemicals which aid the process of digestion. Each enzyme is found in a particular part of the alimentary canal and aids the breakdown of a specific compound (Table 4.2).

Table 4.2 Sources and actions of the digestive enzymes

Compound digested	Enzymes	Source	Product
Starch and dextrins	Ptyalin	Saliva	Maltose
	Amylopsin	Pancreatic juice	Maltose
	Amylase	Intestinal juice	Maltose
Maltose	Maltase	Pancreatic juice	Glucose
	Maltase	Intestinal juice	Glucose
Sucrose	Sucrase	Intestinal juice	Glucose and fructose
Lactose	Lactase	Intestinal juice	Glucose and galactose
Protein	Pepsin	Gastric juice	Proteoses
	Trypsin	Pancreatic juice	Proteoses Peptones Polypeptides
Proteoses Peptones Polypeptides	Trypsin	Pancreatic juice	Aminoacids
Polypeptides	Peptidases	Intestinal juice	Aminoacids

Mouth

Under natural conditions, the pig roots, i.e. it digs up the soil, with its snout and carries the food to its mouth using the pointed lower lip. When pigs are not permitted to root, the food is picked up by means of the teeth, tongue and characteristic movements of the head. In the mouth the food is ground into a pulp by the teeth and mixed with saliva. The enzyme **ptyalin**, in saliva, begins to break down the food. The food is then pushed to the back of the mouth or pharynx and passes down the oesophagus into the stomach.

Stomach

In the stomach the food is churned up by the squeezing action of the muscular walls and the gastric juice is added to it. The gastric juice is produced by the gastric glands in the walls of the stomach and contains mucus, hydrochloric acid and **pepsin**. The mucus lubricates the food. The hydrochloric acid provides an acid environment which prevents the further action of the ptyalin from the saliva, and provides a suitable environment for the action of pepsin.

The pig has a simple stomach which can only digest concentrate feeds such as grains and their by-products, nuts, legume and other seeds and their by-products, fish and mammalian by-products.

Small intestine

In the small intestine, secretions or juices from the pancreas, liver and small intestine complete the process of digestion. The digested food can then diffuse through the walls of the small intestine.

Pancreatic juice contains enzymes for digesting the three major types of food: proteins, carbohydrates and fats. The proteolytic enzymes **trypsin** and **chymotrypsin** partially digest proteins. The enzyme pancreatic **amylase** hydrolyses starch to maltose. The fats are broken down to carboxylic acids and glycerol by the action of lipase.

Bile is secreted continually by the liver. The bile contains no digestive enzyme but is important in digestion because of the presence of the bile salts which emulsify fat globules so that they are more easily digested by lipase.

Intestinal juices contain the following enzymes:

1 several **peptidases** for splitting polypetides into aminoacids:
2 enzymes for splitting the disaccharides, sucrose, maltose, isomaltose and lactose into monosaccharides such as glucose, fructose and galactose;
3 intestinal **lipase** for splitting neutral fats into glycerol and carboxylic acids; and
4 very small amounts of intestinal **amylase** for splitting carbohydrates into disaccharides.

Large intestine

The large intestine is the site for the absorption of water into the body. It secretes mucus to lubricate the undigested food. In response to bacterial infection it secretes large volumes of water and electrolytes. This action dilutes the irritating factors and causes rapid movement of the faeces towards the anus, resulting in diarrhoea.

Absorption

Food absorption does not take place in the mouth and oesophagus. The small intestine is the major site of food absorption.

The mucus membrane of the small intestine is highly modified as an organ of absorption by the presence of a large number of tiny finger-like projections called **villi.** These villi increase the surface area of the intestinal wall and so provide a large surface area for the absorption of food. In pigs the villi are shorter in length than in other farm animals. The absorbed foodstuffs may be transported round the body by the lymph and by the blood portal system.

Fats are absorbed primarily by the lymph. Carbohydrates and products of protein digestion, water and inorganic salts are absorbed primarily by the blood.

Feed composition and the function of foods

Only those feed nutrients that are digested can promote growth, body maintenance and production of milk. The digestion of feed nutrients is influenced by disease, parasites and/or physiological disturbances in the digestive tract. Often such disturbances result in poor growth.

Feeds with high protein and/or high fat content are easily digestible by pigs, but the level of carbohydrates and minerals in the feed does not materially affect digestive efficiency.

In general, the two major components of feed are water and dry matter. The dry matter is composed of proteins, carbohydrates, fats, minerals and vitamins. In Fig. 4.2 the composition of feed is presented diagrammatically.

Energy

All organic nutrients in the feed can be sources of energy for the animal. However, the major sources of energy are fats and carbohydrates. The energy content of the feed is measured in units of heat known as calories (cal) or more commonly, since this is a small unit, as kilocalories (kcal). The correct SI unit for heat is the joule, where 1 cal = 4·186 8 J.

Energy values of feeds are expressed as **digestible energy (DE)**. This is the gross energy of the feed eaten less the energy remaining in the waste products of digestion. Cereal grains such as sorghum, maize, Kafir corn, rice, wheat and millet are high-energy feeds. Cassava, sweet potato, Irish potato and other tubers are medium-energy feeds. Cereal grain by-products such as rice and wheat brans are low-energy feeds.

In some rice-producing countries, like Thailand and the Philippines and Vietnam, rice bran is widely used as an energy feed, while in the South and Central American Tropics, sorghum and the by-products of wheat are the major sources of energy for pigs. In some parts of Africa, root crops such as cassava and yam are used as energy feeds.

The major portion of the mixed diet of pigs consists of energy feeds. To avoid nutrient deficiency the other essential nutrients must be provided in proper proportions to the energy feeds. For instance, growing pigs fed a high-energy diet tend to lower their feed consumption, especially in tropical countries. The high-energy diet will have a low protein content and thus the animal consuming less of this diet is not provided with sufficient protein for growth. In order to produce a large litter and good litter performance, pregnant gilts and sows require less high-energy feeds and more protein in their diet.

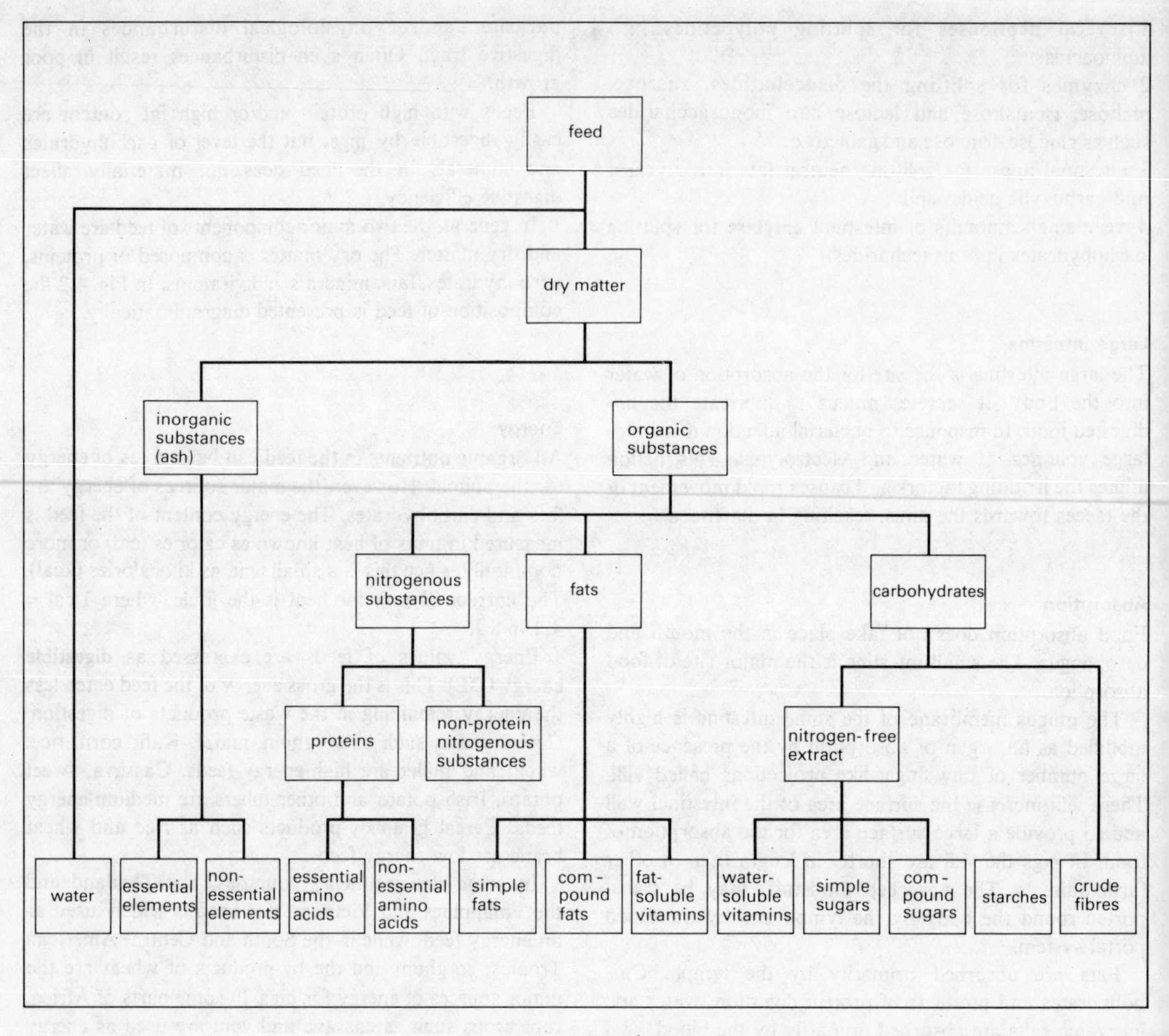

fig. 4.2 Diagrammatic representation of feed composition

Fat

It is recommended that a level of 1·0–1·5 per cent fat should be present in the diet of young pigs. When rice or wheat bran or dried tubers are used as a source of energy, and a solvent-extracted oil meal such as soybean, groundnut or coconut is used as a source of protein, a minimum amount of fat should be added to the ration.

The fat layer of finished pigs fed grated coconut or raw copra (from which the oil has not been extracted) appears soft when compared with that of pigs fed with maize. Pigs fed a high level of cassava meal tend to have slightly firmer backfat than those fed with maize.

A high fat level in the pig's diet improves the feed efficiency. To produce a bacon-type pig, fat or oil should not be given during the growing/finishing stage, i.e. the 60–100 kg liveweight stage.

If fat is used in the diet, an anti-oxidant must be added to prevent the feed from becoming rancid. Butylated hydroxytoluene (BHT) and butylated hydroxyaniline (BHA) are two common anti-oxidants used for this purpose.

Protein

Based on proximate analysis, the protein content of feeds is expressed in terms of crude protein (CP). In order to obtain the CP content the nitrogen content of the feed as analysed is multiplied by 6·25:

$$\text{i.e. } CP = N \times 6{\cdot}25$$

Not all nitrogen present in feeds is true protein or aminoacid. Considerable amounts of non-protein nitrogenous substances such as urea, ammonia and nitrate, are naturally present in feed materials, but they are of no feeding value to pigs. Thus the CP content of a feed does not necessarily provide complete information as to how much protein is available for the pig.

Protein, like carbohydrate and fat, supplies energy. However, the energy yield of 1 g protein is 4 kcal while 1 g of fat provides 9 kcal.

Protein is the most plentiful substance in the animal's body next to water, and is essential in the building up of muscle and other body tissues. These tissues are not permanent but are being constantly replaced throughout the life of the animal and this explains why pigs must be supplied with adequate amounts of protein. In addition, many of the hormones that regulate the metabolic processes of the animal's body, the antibodies which provide immunity from diseases and the catalytic enzymes which speed up chemical reactions inside the body are also proteins.

The great diversity in the functions of proteins are made possible by differences in their molecular structure. The protein molecules are formed from aminoacids which are likened to the letters of the alphabet that comprise the different words of a language. Aminoacids form chains; there are sometimes in a complex protein more than a million aminoacids linked together. Protein is needed in the diet of all classes of pig to supply the aminoacids essential for the maintenance of growth, reproduction and milk production.

Essential aminoacids Arginine, histidine, isoleucine, leucine, lysine, methionine, threonine, phenylalanine, tryptophan, valine. These cannot be synthesised by the pig's body from other aminoacids or sources of nitrogen at all or at a rate rapid enough to produce maximum growth.

Non-essential aminoacids Alanine, aspartic acid, citrulline, cystine, glutamic acid, glycine, proline, hydroxyproline, serine, tyroxine. These can be synthesised by the body, provided there is sufficient nitrogen in the diet.

The animal body cannot store, even for a few hours, unbalanced aminoacid mixtures for protein production. The amount of protein which may be synthesised is limited by the essential aminoacids present in the body and by the total protein supplied in the food.

Pigs can synthesise about 60 per cent of the arginine needed for normal growth. The cystine content in the feed can replace 40 per cent of the required methionine.

Vitamins

Vitamins are organic compounds that are nutritionally essential to pigs but are required in only very small amounts. They are classified according to their solubility. Fat-soluble vitamins include vitamins A, D, E and K which are soluble in fats and solvents such as ether and chloroform. Water-soluble vitamins include the B-complex vitamins and vitamin C (ascorbic acid).

Fat-soluble vitamins

Vitamin A Vitamin A does not exist in plants but its precursor carotene, which does occur in many plants, is converted in the animal's body to vitamin A. One mg of β-carotene is converted into 500 IU of vitamin A in pigs. Concentrate feeds such as grain, with the exception of yellow maize, are not sources of carotene. Green leafy plants such as sweet potato vines and *Leucaena leucocephala* are excellent sources, as is lucerne (alfalfa), when it is available. The leaves of the *L. leucocephala* should not be fed at a rate of more than 4·5 per cent of the ration as this plant contains a toxic substance known as mimosine that causes hair-fall in pigs. Moreover, it exhibits some oestrogenic activity that may cause abortion in pregnant gilts or sows if large quantities are eaten.

Pigs deficient in vitamin A suffer from night blindness. They exhibit a brown exudate around the eyes, become stunted in growth, nervous in disposition and their hind legs become paralysed or stiff. Breeding sows or gilts have poor reproduction and pigs may be born dead, deformed, or sometimes blind.

Vitamin D This vitamin is found in green plants or forages in the form of ergosterol. Ergosterol can be converted into vitamin D by sunlight. The skin of pigs also contains a substance called 7-dehydrocholesterol that is converted into vitamin D by the ultraviolet light in sunshine. Vitamin D_2 (irradiated ergosterol) and vitamin D_3 (irradiated 7-dehydrocholesterol) exhibit the same biological activity in pigs.

As sunlight is abundant in tropical countries vitamin D deficiency is rarely a major problem. Foods deficient in vitamin D include practically all grains and protein food supplements, such as fish meal, copra meal, sesame meal and soybean meal. If all the feed ingredients in pig rations are of plant origin, and the diet is not supplemented with inorganic sources of phosphorus such as bone meal or dicalcium phosphate, a considerable amount of vitamin D will be needed in the diet.

Pigs with vitamin D deficiency possess swollen joints, weak bones and a soft skeleton. They also suffer from paralysis of the hind-quarters.

Vitamin E Maize and other grain feeds are good sources of vitamin E. However, if the major source of energy in pig rations is rice or wheat bran, the pigs are likely to suffer from vitamin E deficiency.

Vitamin E deficiency in pregnant sows or gilts causes high foetal mortality. In suckling pigs it also causes uncoordinated movement of the feet.

Vitamin K is necessary for the prevention of haemorrhagic conditions. It participates indirectly in the blood clotting mechanism by influencing the concentration of prothrombin, proconvertin and thromboplastins in the plasma. Deficiency of the vitamin results in a delayed blood clotting time and haemorrhagic tendency. It is widely distributed in plants and can also be synthesised by intestinal bacteria. Certain conditioning factors can cause vitamin K deficiency such as

1 faulty intestinal synthesis;
2 liver injury;
3 sulphonamide antibiotic therapy;
4 poor intestinal absorption; and
5 obstruction of the bile duct.

Water-soluble vitamins In practice pig rations in tropical countries may be deficient in the B-complex vitamins which include biotin, pantothenic acid, riboflavine, nicotinamide, choline and vitamin B_{12}.

Biotin deficiency causes cracking of the skin and the soles of the pig's feet. The deficiency is, however, only likely to occur when pigs are fed on uncooked eggs, such as those discarded from hatcheries.

Pantothenic acid deficiency causes pigs to develop a stiff-legged walk and their movement becomes uncoordinated. There is a brown colour around their eyes, they possess a poor appetite and their growth is retarded.

Riboflavine deficiency results in stunted growth, a poor haircoat, thick skin, crooked legs and lameness.

Nicotinamide In grain this vitamin occurs in a form that is not available to pigs. Thus, the diet of pigs that are fed solely on grains such as sorghum, maize, oats, rye and rice is likely to be deficient in nicotinamide.

Lack of nicotinamide may bring about a nutritional disease called pellagra. It also causes ulcers on the snout. In pigs, pellagra is characterised by rough ears, skin and haircoat, and a generally poor appearance. The pig also suffers from vomiting and diarrhoea, when the faeces have a putrefied smell.

Choline Feeds containing oil such as fish meal and other animal products are rich in choline. Cereals and cereal by-products contain relatively low amounts of choline.

Pigs deficient in choline exhibit lameness. The legs and joints are stiff and movement is uncoordinated. Growth is retarded on account of the animal's poor appetite.

Vitamin B_{12} This vitamin is plentiful in all protein feeds of animal origin such as fish meal, fish solubles, meat scraps and dairy by-products. Pork producers, however, seldom use feeds of animal origin on account of their high cost. This is especially so in most underdeveloped tropical countries. Therefore, vitamin B_{12} may be lacking in pig rations.

Vitamin B_{12} deficiency retards growth and causes loss of appetite. The animal has an anaemic appearance. Its hindquarters are stiff, and movement is uncoordinated.

Minerals

The mineral elements essential for body function may be classified into two groups; the major elements such as calcium, phosphorus, sodium, potassium, magnesium, and chlorine, and the trace elements, such as iron, iodine, manganese, copper, cobalt, zinc and selenium.

Grains and protein concentrates of plant origin contain phosphorus in an organic form known as phytate phosphorus. Only 50–67 per cent of this plant phosphorus can be utilised by pigs.

When protein concentrates of animal origin are used in the diet, the level of calcium and phosphorus supplementation should be reduced. For instance, there is no need for supplementation with phosphorus if the grower, finisher or breeder rations contain 5 per cent of meat and bone meal. The calcium requirement of a diet containing 5 per cent meat and bone meal may be satisfied by adding 0·5 per cent limestone. Likewise, the calcium requirement of a diet that contains 5 per cent fish or shrimp meal, may be satisfied by adding 1 per cent oyster shell or limestone.

Iron is necessary in the formation of haemoglobin, an iron compound which enables the blood to carry oxygen for cellular respiration. Iron is also important for antibody production and for certain enzyme systems, for example that catalysing the conversion of β-carotene, the precursor of vitamin A, to vitamin A.

Copper improves growth rate and feed conversion efficiency in young pigs. Although it does not contribute to red blood cells, copper is necessary for haemoglobin formation together with iron and vitamin B_{12}. Copper is also essential in enzyme systems, hair development, pigmentation, bone development, reproduction and lactation.

Mineral salt supplements in pig rations should range from 0·25 to 0·5 per cent of the ration depending on the age of the pig. Ample amounts of trace elements are normally supplied by the feeds used in most rations. However, if the trace element deficiencies are suspected, vitamin premixes containing trace elements and antibiotics are commercially available in many countries.

Water

About 60–70 per cent of the total body weight is comprised of water, which is second only to oxygen in importance for the maintenance of life. All the chemical reactions in the body take place in the presence of water. It acts as a solvent for the products of digestion, as a lubricant for moving parts and as a regulator of body temperature. Blood is 90 per cent water and urine is 97 per cent water. It is also important for the proper elimination of body waste products. Sources of water for the body are:

1 water as such (as a drink);
2 water in food and body tissues; and
3 metabolic water provided by the combustion of foodstuffs such as carbohydrates, protein and fat.

In tropical countries pigs should be given 4–5 kg of water for every 1 kg of dry feed. Lactating sows should always have free access to drinking water in order to produce an adequate amount of milk for their nursing litters.

Pre-treatment of pig feeds

Pig feeds generally undergo some kind of processing or treatment before they are mixed into a diet or feed. For instance, soybean oil meal is a by-product of soybean after the oil has been extracted. This pre-treatment of the soybean meal affects its digestibility by the pig. The solvent-extracted soybean meal is better utilised by pigs than the expeller or heat-extracted meal. Moderate drying of protein feeds promotes their feeding value. When heat-treated at a temperature of not more than 100°C for 12 minutes, soyin, a toxic substance in soybean, is eliminated. On the other hand, overheating of soybean lowers its feeding value.

Grinding of cereal grains increases the animal's efficiency for utilizing them as energy sources. Unground maize has a digestibility of only 70 per cent, whereas for ground maize the figure is 86 per cent. However, very finely ground maize and maize bran can cause gastric ulcers.

Cooking maize or any other feed does not apparently improve their digestibility for pigs but may have other advantages. Although the boiling of potatoes or sweet potatoes does not affect digestibility it improves the growth rate of pigs. The favourable effect of cooking root crops on the growth of pigs is apparently due to the destruction of fungus or moulds that may be present on them.

Neither soaking nor wetting the feed has any effect on digestibility. Thus, from this standpoint, dry or wet-feeding is immaterial. In practice in the Tropics, however, the feeding of a wet mash can improve the appetite of the pigs, provided that no more than twice as much water is used as solid feed.

Feeding management in pigs

The daily nutritional requirements of pigs are highest during the early phase of their life and gradually decrease as they approach maturity. During early life, the natural resistance of the pig is not yet fully developed. It therefore needs maximum assistance to help it carry on the normal biological functions of the body. The early phase of the pig's life as a nursing baby pig coincides with the lactation stage of the mature sow. During the nursing or lactation stage of the sow, her body requires maximum nutrients because milk production places a heavy drain on body reserves. In the Tropics, temperature adversely affects the milk secretion of lactating sows. The nutrient allowance, therefore, during the lactation or nursing stage, should be considerably above maintenance requirements.

As can be seen from Fig. 4.3, the wide sections of the shaded bar denote the most critical period and the narrow sections of the shaded bar represent the less critical period in the life of the pig. During the less critical period, disease resistance is well developed, and nutritional demands are lower.

Flushing and gestation phase

During the pre-gestation stage, when the litter has just been weaned from the dam, the sow is 'flushed' for a period to condition her for the coming breeding season. Flushing consists of increasing the amount of high-energy feeds given to the sow. The high-energy ration may also include an increased protein content. Improved nutrition during flushing of the sow or gilt enhances the capacity of her ovaries to produce additional eggs during oestrous.

A week before and after mating the protein intake is further increased in order to replace the tissue proteins that are used during ovulation. Subsequently protein and energy feed intake is reduced to a minimum from one week after mating until one week before the scheduled farrowing date. Vitamin A should, however, be provided at a maximum level corresponding approximately to 4 100 IU per kg of the ration. The addition of an antibiotic to the diet during gestation does not increase the litter size at birth but does improve the birth weight and livability of the pigs. From an economic and practical viewpoint, antibiotics should not be added to the diet during the gestation phase as their addition does not essentially affect the number of eggs fertilised and hence the

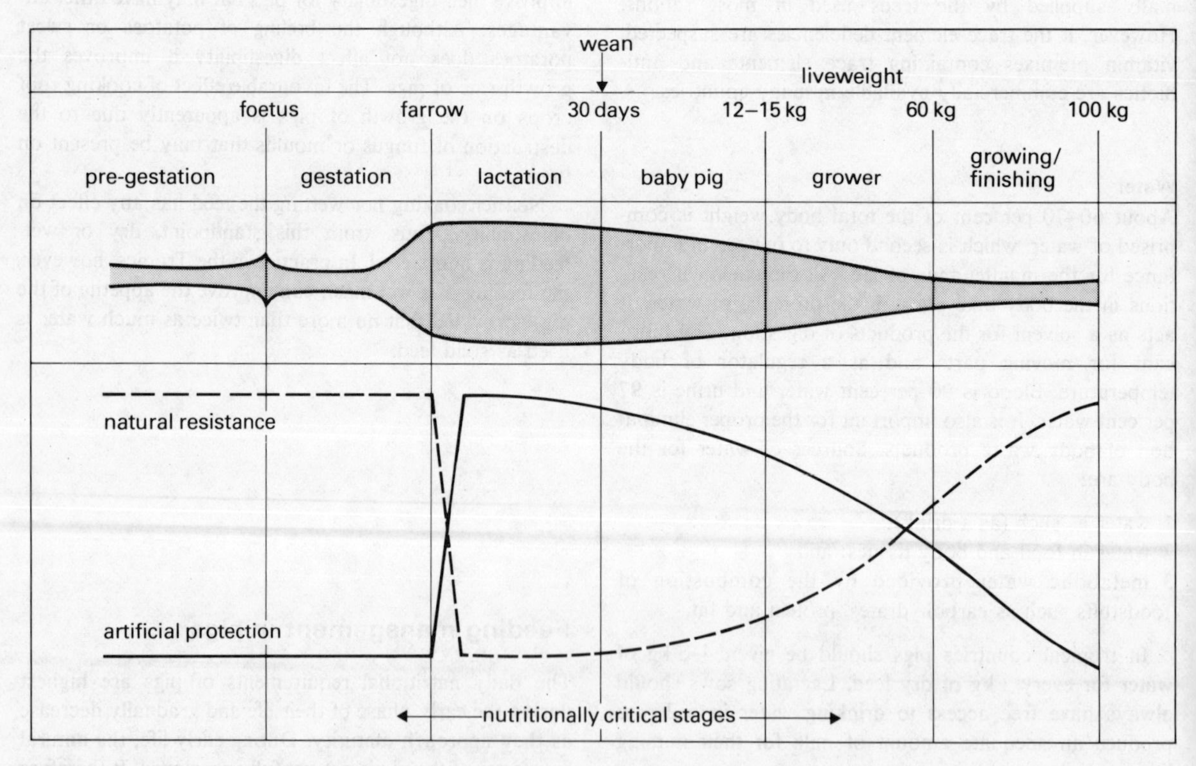

fig. 4.3 Modern phase feeding in tropical pig production

size of the litter. However, if the purpose is to increase the birthweight of the litter, then the use of antibiotics may be justified.

It should also be mentioned that during the gestation phase, supplementing the diet with hormones, such as diethylstilboestrol, may cause abortion and can only be detrimental or harmful.

Adequate feeding of the boar is as important as that of the sow. An active boar should be maintained on a high protein diet together with an adequate vitamin A supplement. Vitamin A or β-carotene can be provided by a daily allowance of fresh succulent forage such as sweet potato or kang-kong (*Ipomoea reptans*).

Farrowing phase

A week before the scheduled farrowing date, feed is increased. This provides the sow with food reserves necessary to enable her to meet the body stress that comes with farrowing. A few days after farrowing the sow may not have much of an appetite. When her appetite returns she should be given maximum feed to enable her to rebuild the body reserves that she used up during farrowing.

Lactation/baby pig phase

It is the lactation/baby pig stage in the life cycle of the pig that is considered the most critical period. This is because nursing pigs are low in natural body resistance. At this stage baby pigs need maximum artificial protection and lactating sows need an increase in nutritional supply.

Newborn pigs need antibodies. They normally obtain these from the colostrum or first milk of the sow within a few hours after birth and not prenatally. The absorption of antibodies decreases to insignificant amounts in baby pigs that are more than 24 hours old. While there is a decline in antibody absorption during this stage, environmental stress is always present and is frequently accompanied by some disease agent. This is especially so in the Tropics. According to studies, the 'alarming stimulus' or stress factors such as high temperature, humidity, dampness and cold nights, etc., set up a chain reaction that creates a need for increased amounts of certain specific nutrients. Several vitamins, the majority of which are water-soluble fractions of the vitamin B-complex, are considered important for reducing the severity of this stress syndrome. These vitamins also hasten the recovery of the pigs from the ill effects of stress.

For reasons stated above, fortification of the baby pig's diet with vitamins and trace elements plays an important role in helping the body to adjust to the accelerated cell activity which occurs during the early phase of life.

In addition, growth rate is rapid at this stage and body cells are multiplying fast. The animal needs a high level of vitamins and trace elements to support rapid cell multiplication, which results in efficient and economical weight gain. It has been reported that it is muscle cells that are laid down during this early phase of life. A small-sized cell has a large surface area in relation to its volume. As a result, a large number of small cells have a higher metabolic rate than a few larger-sized cells. It is during the lactation/baby pig stage that pigs are most efficient in utilising feed nutrients that result in growth or liveweight gain.

Grower phase

The critical period in the life of the pig continues until it reaches the age of puberty or attains a liveweight of 60 kg. After this grower stage, the pig is considered to be in the self-inhibiting phase of growth or in the grower/finisher stage. Although the cells still continue to multiply, multiplication is not as rapid as in the earlier phases. The natural resistance of the pig is well developed and pigs require the minimum protection from environmental stress. The nutritional requirements, therefore, start to decline during this stage and continue to decrease until the pig reaches 100 kg liveweight.

Growing/finishing phase

When a pig has attained 60 kg liveweight its rate of growth and feed conversion efficiency decreases, so that feed requirements for production are decreased but nutrient requirements for maintenance are increased. This is because the tissue cells increase in size rather than in number and, therefore, the rate of metabolic exchange is less than during the earlier phases of life.

It is also at the growing/finishing stage that replacement gilts are selected as future breeders in the herd. As the nutritional requirements of growing/finishing pigs are lower than those of replacement gilts, the latter should be fed rather better.

Table 4.3 Calculation of protein content and digestible energy content of pig feeds

Protein content:

Feed to be used	CP (%)	Quantity of feed (kg)	Estimated CP content (kg)	
Sorghum	8·8	84·0	(84 × 0·088)	7·39
Fish meal	65·0	13·0	(13 × 0·65)	8·45
Mixed feed		97·0		15·84

Digestible energy content:

Feed to be used	DE (kcal per kg)	Quantity of feed (kg)	Estimated DE content (kcal per kg)	
Sorghum	3 414	84·0	(84 × 34·14)	2 868
Fish meal	2 994	13·0	(13 × 29·94)	389
Mixed feed		97·0		3 257

Compounding a pig diet

A simple grower diet using sorghum as a source of energy and fish meal as a source of protein may be formulated in several ways as in Tables 4.3, 4.4 and 4.5.

Assuming the crude protein (CP) content of sorghum and fish meal to be 8·8 and 65 per cent respectively, the protein and digestible energy content of the ration is calculated as shown in Table 4.3.

In a similar manner the calcium and phosphorus contents may be calculated as shown in Table 4.4.

Phosphorus in sorghum is in the form of phytin phosphorus and only 67 per cent of this can be utilised by pigs. Therefore, the available phosphorus content of sorghum is 0·67 × 0·311 or 0·208 per cent and adding the phosphorus from fish meal the total available phosphorus is 0·520.

The ration will require addition of salt and a vitamin/mineral premix. As the calcium and phosphate contents are not high, bone meal should be added. The nutrient contents of the feed are shown in Table 4.5. Under tropical conditions this level of energy (3 257 kcal per kilogram) for a growing pig is adequate for satisfactory growth. It will be seen that the calculated nutrient content of this mix should be sufficient for the nutritional needs of growing pigs of 20 to 50 kg liveweight.

Table 4.4 Calculation of calcium and phosphorus contents of pig feeds

Calcium content: Feed to be used	Calcium (%)	Quantity of feed (kg)	Estimated calcium content (%)
Sorghum	0·02	84·0	(84 × 0·000 2) 0·017
Fish meal	4·50	13·0	(13 × 0·045) 0·585
Mixed feed		97·0	0·602

Phosphorus content: Feed to be used	Phosphorus (%)	Quantity of feed (kg)	Estimated phosphorus content (%)
Sorghum	0·37	84·0	(84 × 0·003 7) 0·311
Fish meal	2·40	13·0	(13 × 0·024) 0·312
Mixed feed		97·0	0·623

Table 4.5 Nutrient content of a grower feed

Ingredient	Quantity (kg)	CP (%)	DE (kcal per kg)	Ca (%)	P (%)
Sorghum	84·0	7·39	2 868	0·017	0·201
Fish meal	13·0	8 45	389	0·585	0·312
Bone meal	2·0			0·048	0·024
Salt	0·5				
Vitamin/mineral premix	0·5				
Mix	100·0	15·84	3 257	0·650	0·544
Recommended nutrient requirements (see Table 4.2)		16·00	3 500	0·650	0·500

Feed formulations for pigs

Suggested feed formulations for the different stages in the life cycle of pigs are presented in Tables 4.6 to 4.10. These rations are generally intended for restricted hand-feeding, except in the case of early-weaned pigs which should be fed as much as they wish to eat. Growing pigs from weaning until they attain 60 kg may be put on self-feeding for 1 hour; once in the morning and again in the afternoon. Gilts and sows should be on rationed and restricted feeding; they should be given 1·5 kg in the morning and another 1·5 kg in the afternoon. Lactating sows should also be rationed, though the total feed intake should be increased to 4 kg daily. Active breeding boars should be given the same rations as sows on restricted feeding.

To produce bacon-type pigs feeding management must be changed at 60 kg and the new ration continued until they attain 90 to 100 kg. If the pigs are well fed during their growing stage, bone and muscle tissues represent most of their growth. If they are then rationed or put on restricted feeding after they reach 60 kg liveweight, excess fat growth is prevented and muscle growth continues, in this way a bacon carcass is produced. The tendency for finishing pigs to become overfat must be controlled by increasing the bulk of the ration by the use of feeds high in fibre content such as rice and wheat brans, oats and barley.

Table 4.6 Sow pre-gestation and gestation rations

Feed	1	2	3	4	5
Ground maize or sorghum	26·5		36·5	41·5	25·0
Rice or wheat bran	26·0	25·0	25·5		50·0
Oats				20·5	
Ground maize with cob		34·0			
Rice or wheat middlings	10·0				
Molasses (cane)	5·0	5·0			
Meat and bone scraps	5·0	5·0			
Soybean oil meal	2·5	10·0	6·0	10·0	11·0
Fish solubles (50% solids)	2·5				
Fish meal			2·0		
Maize gluten feed					8·0
Distillers dried solubles	2·5				
Leucaena leucocephala leaf-meal					3·0
Alfalfa (lucerne) meal (17% dehydrated)	19·0	20·0	28·0	25·0	
Salt	0·5	0·5	0·6	0·5	0·5
Oyster shell			0·5		2·0
Bone meal			0·4	2·0	
Vitamin/mineral premix commercial mixture	0·5	0·5	0·5	0·5	0·5
Total	100·0	100·0	100·0	100·0	100·0

Table 4.7 Farrowing/lactating rations

Feed ingredient	1	2	3	4	5
Ground maize or sorghum	48·3	71·3	40·2	55·0	51·0
Rice or wheat bran	15·0		50·0	20·0	15·0
Oats (finely ground)	5·0		15·0		5·0
Molasses			8·0		
Meat and bone scraps	5·0		5·0		5·0
Soybean oil meal	10·0	14·5	10·0	18·0	12·2
Fish solubles	2·5		2·5		
Fish meal		5·0			
Maize gluten feed	2·5			4·0	
Copra (coconut) meal			2·5		
Alfalfa (lucerne) meal	10·0	10·0	10·0		10·0
Oyster shell	0·2			1·1	0·1
Dicalcium phosphate	0·5		0·8		0·7
Bone meal		3·2		0·9	
Salt	0·5	0·5	0·5	0·5	0·5
Vitamin/mineral premix	0·5	0·5	0·5	0·5	0·5
Total	100·0	100·0	100·0	100·0	100·0

Table 4.8 Starter rations for early-weaned (35 day) pigs

Feed ingredient	1	2	3	4	5
Ground maize or sorghum	30·0	51·0	46·0	59·0	48·0
Fine rice or wheat bran		13·0	15·1	15·0	17·5
Rolled oats	30·0				
Sugar			8·0		2·5
Molasses	5·5	3·5			2·5
Soybean oil meal	7·5	16·0	25·7		21·5
Meat meal				10·0	
Fish meal	8·0	4·0		5·0	
Dried skimmilk	10·0	10·0		10·0	5·0
Alfalfa (lucerne) meal	4·0				
Dried brewers' yeast	3·8				
Salt	0·5	0·5	0·5	0·5	0·5
Ground oyster shell	0·2	0·5	2·0		1·0
Bone meal		0·5	2·2		1·0
Vitamin/mineral premix	0·5	0·5	0·5	0·5	0·5
Total	100·0	100·0	100·0	100·0	100·0

Table 4.9 Grower pig rations. Weaning to 60 kg liveweight

Feed ingredient	1	2	3	4
Yellow or white maize (ground)	55·0		68·5	
Sorghum or milo		64·4		78·9
Fine rice bran	12·0	12·5		
Rice or wheat middlings	10·0			
Meat and bone scraps	2·5	2·5		2·5
Soybean oil meal	18·5	18·5	21·0	16·1
Fish soluble (50% solids)			2·5	
Dried wheat			2·5	
Distillers' dried solubles			2·5	
Salt	0·5	0·5	0·5	0·5
Ground oyster shell	0·9	0·6	1·0	0·5
Dicalcium phosphate	0·1	0·5	1·0	1·0
Vitamin/mineral premix	0·5	0·5	0·5	0·5
Total	100·0	100·0	100·0	100·0

Table 4.10 Growing/finishing rations for bacon-type pigs (60–100 kg)

Feed ingredient	1	2	3	4
Fine rice or wheat bran	35·0	31·0	45·0	41·0
Yellow or white maize (ground)		45·0	36·0	
Sorghum or milo	44·0			35·0
Maize gluten feed		7·0		
Soybean oil meal	10·0	10·0	9·0	9·0
Copra (coconut) meal	4·0			6·0
Fish meal	2·0	2·0	2·0	
Molasses	2·0	2·0	5·0	6·0
Ground oyster shell (limestone)	1·7	1·5	1·6	1·8
Bone meal	0·3	0·5	0·4	0·2
Salt	0·5	0·5	0·5	0·5
Vitamin/mineral premix	0·5	0·5	0·5	0·5
Total	100·0	100·0	100·0	100·0

Common feeds for pigs in the Tropics

A well-balanced pig ration should contain the necessary amounts of energy feed, protein, minerals and vitamins. The energy in the diet may be supplied by cereal grains, cereal by-products, root crops, molasses, sugar, fats, fruits or fruit by-products. The protein feed supplements should be oilseed cakes, legumes, nuts, slaughter house by-products or fish by-products.

Minerals and vitamins are obtained from the feeds used in the ration, or from commercially prepared supplements. Sources of micronutrients will be discussed later.

Energy feeds 1 Cereal grains

Cereal grains are considered to be first-class energy feeds for pigs. They can be produced easily in the Tropics and can be made available throughout the year provided there are adequate storage facilities.

Maize Yellow and white maize are excellent sources of energy. They can be used at the rate of 85 per cent in the diet of growing pigs. For pregnant sows, maize can be used at a minimum rate of 25 per cent. Yellow maize is rich in β-carotene but is a poor source of vitamins B_{12} and D.

Wheat This is a high-energy feed commonly used in South America and Australia and available in montane tropical regions. The feeding value of wheat is equal or slightly inferior to that of maize. For the best feeding results wheat should not be ground too finely.

Rice This cereal is somewhat similar to wheat in feeding value. The rice hull is very high in fibre and lignin. Whole rice has about 75 to 85 per cent of the feeding value of maize. It can replace 50 per cent of maize in the pig's diet and still produce good results.

Barley Although barley has slightly less energy value than wheat it is still a good source. It is commonly used as a feed in Africa, Europe and Western Australia. Barley has about 90 per cent of the feeding value of maize. It should be ground before it is used.

Sorghum Grain sorghum has about 95 per cent of the feeding value of maize. Rolled sorghum has a better digestibility in organic matter, protein and carbohydrate fraction than whole-grain sorghum. A growing/finishing ration may contain as much as 85 per cent of rolled sorghum.

Rye This grain has about 90 per cent of the feeding value of maize. It is not as palatable as other cereals so it should be ground and mixed with other grains. For growing/finishing pigs, it can be added at a level of not more than 50 per cent of the diet. Rye is seldom available in the Tropics.

Oats This cereal has 60 to 90 per cent of the feeding value of maize, but its high fibre content restricts its use to up to 30 per cent of the ration for growing/finishing pigs and for brood sows.

Rice or wheat shorts These are by-products of the milling of rice or wheat, but they do not contain hulls. Their feeding value is 90 to 115 per cent of that of maize. They can be included in rations for growing/finishing pigs at the rate of 60 per cent of the total.

Energy feeds 2 Root crops and miscellaneous feeds

Cassava (*Manihot esculenta*) The root of cassava, also known as manioc, yuca and tapioca is used extensively as a feed for livestock. It may be used as either fresh or dried roots for energy feed in pig rations. Dried cassava or cassava flour has practically the same feed value as maize. Cassava leaves are used in the form of leaf meal as a protein source in livestock and poultry rations.

Extensive studies on the use of cassava roots as a pig feed have been carried out at the International Centre of Tropical Agriculture (CIAT) in Colombia, in Venezuela, in Brazil, in Malaysia, in the Philippines and in Singapore.

In general, results from these different studies conducted in Southeast Asia and South America indicate that the inclusion of over 50 per cent cassava meal in the ration decreases liveweight gain and the feed conversion efficiency of the pig. However, when methionine is supplemented at the 0·2 per cent level in the diet, both feed efficiency and weight gain are improved. It has also been reported that the addition of palm oil and glucose in the diet, improves the daily weight gain and feed conversion efficiency of pigs fed cassava. Supplementation with palm oil enables pigs to utilise higher levels of methionine in the diet. On the other hand, the addition of glucose to the diet is considered to reduce the hydrocyanic acid content of cassava by forming gluconohydrin and thereby improving its palatability. This results in increased feed intake. It was also observed that the overall performance of the pigs including their health and carcass quality are not significantly impaired by an increased level of cassava meal, provided that the feeds are pelleted and carefully balanced with regard to aminoacid content.

Extensive feeding trials on the use of cassava root soilage as a component of pig rations have also been conducted at CIAT. The low protein level of the cassava soilage presents a problem and attempts to raise this by microbial action have not been successful.

In some countries cassava roots are cooked to destroy the poisonous cyanogenetic glucosides found in the skins of some varieties.

Sweet potato (*Ipomoea batatas*) Dried sweet potato has 90 per cent of the feeding value of maize. It can replace 30 to 50 per cent of the grain in pig rations. In Taiwan, sweet potato and cassava are used to replace 40 per cent of the maize in growing and growing/finishing diets.

Irish potato When potato is mixed with grain in the ratio of 3:1 the feeding value is only 30 per cent of that of maize. Dried potato should constitute no more than 30 per cent of the pig ration.

Molasses Molasses has a high feeding value when used to replace the less palatable grains, but should be used with care as it may cause digestive troubles. A level of less than 30 per cent of the ration is generally recommended for fattening pigs and less than 5 per cent for young pigs. In Cuba, feeding growing pigs with 30 per cent cane molasses in the diet resulted in a poor growth rate. However, when molasses was combined with sugar, the feed efficiency was greatly improved. In Colombia growing/finishing pigs on a 13 per cent protein ration containing 30 per cent molasses showed the highest daily weight gain and improved feed efficiency.

Fermented corn cobs, seaweed, rice straw and banana stalks Fermented feeds when fed to pigs do not show the desirable effects on growth and feed efficiency produced by unfermented feeds and feed concentrates. Fermentation, however, improves the palatability and aroma of feeds, which in turn increases the overall feed consumption. Fermented corn cobs and banana stalks provide the bulk and aroma of the diet. Fermented seaweeds and rice straw can replace 5–10 per cent of maize in growing/finishing rations.

Citrus molasses Citrus molasses has the same feeding value as cane molasses. However, citrus molasses is used at a lower level since it has a bitter taste and must be mixed with either cane molasses or sugar.

Chick pea (*Cicer arietinum*) In Mexico, chick pea has been used in pig rations as a source of energy and protein. A diet containing more than 80 per cent chick pea did not appreciably affect liveweight gain but produced a slight improvement in feed efficiency. When 0·3 per cent methionine was added to the diet, feed efficiency improved but not liveweight gain.

Sugar The most practical and economical use of sugar is to improve the palatability of baby pig starter diets. When used as a partial substitute at the 5 per cent level in the ration, the feeding value of sugar is 112 per cent that of maize.

Energy feeds 3 Cereal and fruit by-products

Wheat bran The feeding value of wheat bran is 67 per cent that of maize. However, on account of its high fibre content wheat bran should only be used for the feeding of growing/finishing pigs and brood sows. In some tropical countries where maize is expensive, first-class wheat bran is used at levels of up to 40 per cent in rations for brood sows.

Rice bran This bran has about the same feeding value as wheat bran. It is rich in thiamine and nicotinamide. However, it cannot be used as a total substitute for maize because of its high fibre content and its laxative effect, especially for young pigs. In the Philippines first-class rice bran is used at a level of up to 50 per cent of the ration for growing and finishing pigs.

Citrus pulp This pulp has 47 per cent the feeding value of maize when used at a level of 10 per cent of the ration for growing pigs. For satisfactory results, however, it should never be used at a level of more than 5 per cent in the ration.

Maize bran Maize bran has 86 per cent of the feeding value of maize. High levels of maize bran in the rations for growing pigs may cause gastric ulcers. For satisfactory results it should be used in combination with other cereal grains.

Table 4.11 presents the relative feeding values of some selected energy feeds using maize as an index. For practical and economic reasons, sugar, dried whey, animal fats and plant oils should not totally replace maize in pig rations. Molasses is cheap in some countries but its high potassium content makes it unsuitable as a total replacement for maize; it can cause diarrhoea if given at the rate of more than 7 per cent in the ration; and see above.

Table 4.11 The relative feeding values of selected energy feeds

Energy feed		kg – 1 kg maize
Yellow maize, ground		1·00
Complete substitutes		
Cassava (*Manihot esculenta*) dried flour		1·05
Sweet potato (*Ipomoea batatas*), dried		1·10
Elephant yam (*Amorphophallus campanulatus*)		2·90
Sugar		0·89
Grated coconut meat		0·75
Paddy rice (*Oryza sativa*)		1·20
Grain sorghum (milo)		1·05
Wheat grain		1·05
Wheat bran		1·50
Rice bran (coarse)		2·63
Rice bran (fine)		1·50
Partial substitutes	**Percentage level to replace maize**	
Molasses (sugarcane)	5–7	1·69
Spent brewers' grains	50	1·00
Sugar	5	0·89
Grated coconut meat	50	0·65
Dried whey	5	1·11
Oats (ground)	25	1·22
Tallow	5	0·41
Coconut oil	5	0·43
Lard	5	0·43

Protein feed supplements

Soybean oil meal Soybean oil meal is one of the most common feed ingredients used as a protein supplement for young growing pigs and brood sows. The solvent-extracted product is a better quality soybean meal than the expeller-processed product. Soybean oil meal is the best quality protein feed of plant origin available.

Meat meal The quality of meat meal varies considerably. Many so-called meat meals are really meat and bone meals. The feeding value of meat meal is 89 per cent that of soybean oil meal. Its use in the ration may range from 5–10 per cent depending on the cost and availability of protein feeds in the country.

Fish meal The most common fish meals used in the Tropics are those made from the manhaden and the herring. The herring fish meal is better than manhaden. The recommended levels to use range from 2–10 per cent of the ration. Fish meal should never be included in the finishing rations.

Whale meal Whale meal is a good source of protein but it is unpalatable. If combined with other protein feed supplements it will produce satisfactory results. It has been commonly used in Western Australia for growing pigs but it is not likely to be freely available in the future.

Shrimp meal The feeding value of shrimp meal is 54 per cent that of soybean oil meal. A combination of shrimp and soybean oil meals produces better results than the use of shrimp meal alone.

Blood meal The feeding value of blood meal is 75 per cent that of soybean oil meal. Blood meal is available and cheap in some tropical countries of South America. Its use in pig rations should not exceed 5 per cent of the total.

African snail (*Achatina fulica*) The African snail was introduced in the Philippines during World War II, and has developed into a serious pest of rice, bananas and other plants. When it is cooked for 30 minutes, removed from its shell and used in pig rations it produces satisfactory results without ill-effects on the pigs. It has been used for breeding sows at a level as high as 50 per cent in a diet that included rice bran and fresh forage, and the animals had a satisfactory production performance.

African snail meal has also been used as a substitute for fish meal in chick's ration. The results indicate that African snail meal can satisfactorily replace fish meal. The snail is available in many other tropical countries in Southeast Asia and in West Africa.

Skim milk The feeding value of dried skim milk is 101 per cent that of soybean oil meal. It is most useful for baby pig starter rations at levels of 10–30 per cent. It is not normally economic to use it for brood sow and finishing hog rations.

Linseed oil meal Linseed oil meal has a feeding value of 96 per cent that of soybean oil meal. However, its use should be limited to a maximum of 8 per cent of the total ration. For growing pigs and brood sows it should not make up more than 5 per cent of the total ration.

Safflower meal (*Carthamus tinctorius*) The feeding value of safflower meal is 65 per cent that of soybean oil meal, but its use is very limited. Used at the rate of 12·5 per cent of the ration it does not have any beneficial effect on liveweight gain and feed efficiency. When added at the rate of 25 per cent of the ration, it causes a depression in growth. It has a low content of lysine and methionine.

Sesame (*Sesamum indicum*) oil meal The feeding value of sesame oil meal is 89 per cent that of soybean meal, but its use is limited to 2–5 per cent of pig rations. In South America it is only used for growing/finishing pig rations.

Cotton seed meal Some processed meals contain a toxic substance known as gossypol that limits their use in pig rations. In Colombia, cotton seed meal contains only very small amounts of gossypol. It should not be used at a rate of more than 10 per cent in pig rations.

Coconut oil (copra) meal The feeding value of coconut oil meal is 80–90 per cent that of soybean oil meal. However, its use in pig rations should not exceed 30 per cent of the total.

Groundnut (peanut) meal Although groundnut meal has been reported to have the same feeding value as soybean oil meal its use is limited on account of the possibility of aflatoxin poisoning. The toxin is produced by fungi growing on the groundnut, after harvesting. In Malaysia and elsewhere, aflatoxin poisoning has been reported to be the cause of the death of young pigs fed groundnut meal. Groundnut meal also becomes rancid when stored for a long period. It can be satisfactorily used at a 5 per cent level in growing pig rations in combination with other protein feeds.

Feather meal Hydrolysed feather meal can satisfactorily constitute 5 per cent of the diet for growing/finishing pigs. In a Florida study, it was reported that feather meal was unsatisfactory as the only source of protein but when used at the 5 per cent level in a maize/soybean meal diet, good growth was obtained.

Maize (corn) by-products Maize germ meal and maize gluten feed have about 95 per cent the feeding value of soybean oil meal. Maize gluten meal and dried distillers' solubles are of the same feeding value compared to soybean oil meal. Dried distillers' soluble products are good sources of B-complex vitamins. However, the use of either feed in a pig ration should not exceed 5 per cent. It is better to use them in combination with other protein feeds.

Vitamin feed supplements

Feeds contain naturally occurring vitamins but they are usually not present in sufficient quantity to satisfy the requirements of pigs. Therefore, even pigs receiving well formulated rations containing adequate energy, protein and minerals still need good pasture feeding to supplement the vitamins in their ration. However, a problem in any pasture feeding system is that parasites, especially kidney worms, are very common in the Tropics. Under these conditions, vitamin preparations in the form of premixes, should be used to supplement the rations of indoor fed pigs.

Fat- and water-soluble vitamins occur naturally in common pig feed materials as shown below.

Fat-soluble vitamins Vitamin A is present in yellow maize, alfalfa meal, *Leucaena leucocephala* leaf meal, fish liver oil and good pasture. Ramie (*Boehmeria nivea*) leaves, sweet potato leaves and kang kong (*Ipomoea reptans*) are very rich sources of carotene. Vitamin D_2 is obtained from irradiated yeast, sun-cured legume meals or sunshine. Vitamin E is present in maize and wheat germ meal and cereal grains. Vitamin K is present in alfalfa meal or pasture.

Water-soluble vitamins Thiamine, riboflavin, pantothenic acid, nicotinamide, pyridoxine, and B_{12} are present in fish solubles, rice bran, distillers' solubles, milk by-products, brewers yeast, fermentation products and good pasture.

Mineral supplements

Major elements Calcium, phosphorus, sodium and chlorine are supplied by limestone, steamed bone meal, dicalcium phosphate and common salt. Potassium and magnesium are supplied by grains or roughages.

Trace elements are commercially available as mineral premixes or vitamin/mineral premixes.

Feed additives, such as antibiotics, additional copper, arsenical compounds, surfactants, hormones, etc., are growth stimulants and can be supplied either individually or with the vitamin/mineral premix.

Climatic environment and nutrition

The high environmental temperature and high relative humidity in many tropical countries, especially in Southeast Asia, the Pacific and Caribbean regions, West Africa and South and Central America, can greatly affect the nutritional requirements of pigs. High ambient temperatures decrease the appetite of the pig. A poor appetite decreases the feed intake resulting in poor growth performance of the animal.

Under humid tropical climatic conditions, pig diets should not contain high levels of energy feeds because these diets can result in protein deficiency. High-energy feeds, e.g., maize, sorghum, barley, root crops, are generally poor sources of protein. Therefore, when these feeds are used in large amounts to increase the energy level of rations, the protein content of the diet may be reduced to a level below normal requirements. In order to obtain normal growth performance of pigs raised in a tropical environment, the level of energy in their diets should be slightly lower than that of diets fed to pigs in a temperate climate.

Low-energy rations allow additional protein feed supplements to be included in the diet. In general, pigs provided with a low-energy diet increase their feed intake in order to satisfy their energy needs. When an animal increases the feed intake, it will make adjustments to satisfy its protein needs as long as there is an adequate supply of protein in the diet. It is important to stress, however, that with low-energy diets feed conversion performance may be decreased, while high-energy diets generally improve the feed conversion efficiency. In both cases, however, the liveweight gains of the animals may not necessarily be affected, provided other limiting nutritional factors, such as aminoacids, vitamins, and minerals, are in adequate supply.

In South Africa, it has been found that on an extremely hot day a sudden drop of temperature in the evening affects the daily feed intake and nutrient requirements of the pigs. Vitamin A and some of the B-complex vitamins, such as pantothenic acid and pyridoxine, have been found to be important for pigs raised under conditions where sudden changes in environmental temperature occur frequently.

Further reading

Beames, R. M. (1969). A comparison of the digestibility by pigs of whole and rolled sorghum grain fed restricted or *ad libitum. Australia J. Exp. Agric.*, **9**, 127–130.

Becker, D. E., Jensen, A. H. and Harmon, B. G. (1963). *Balancing swine rations.* The Illinois System of Swine Nutrition, Univ. of Illinois, College of Agric. Coop. Ext. Serv. Cir. 866.

Castillio, L. S., Gerpacio, A. L., Gloria, L. A., Skinner, E. C., Arganosa, V. G., Perez, C. B. and Garcia, G. V. (1966). Fermented corn cobs, seaweeds, rice straw and banana trunks for fattening pigs. *Phil. Agric.*, **49,** 799–815.

Catron, D. V., Speer, V. C., Hays, V. W., Jones, J. D. and Dias, F. (1959). *Applied life cycle swine nutrition.* Iowa State Univ. Coop. Ext. Serv. AH 778.

Comb, G. E. (1965). Swine production in Florida. *Florida Dept. Agric. Bull., No. 21*, Gainesville: Florida.

Eusebio, J. A., Acosta, D., Calilung, V. and Alcantara, P. F. (1968). Case report: gastric ulcers in pigs fed high levels of bran. *Phil. Agric.*, **52,** 233–240.

Eusebio, J. A. (1969). *The science and practice of swine production.* UPCA Textbook Board: Los Baños, Philippines.

Eusebio, J. A., Liebholz, J., Mendoza, R. B., Eusebio, E. C., Caparas, L., Supnet, M. G., Gatmaitan, O. M. and Momoñgan, V. (1977). Nutrient requirements of swine under tropical humid conditions. *National Science Development Board—National Research Council of the Philippines Tech. Bull.*, **2** (2) pp. 1–29: Los Baños, Philippines.

French, M. H. (1963). Efficiency of feed conversion by pigs in the tropics. *Agron. Trop. Venezuela*, **12,** 165.

Gallardo, C. R. (1960). A study of the effect of varying amounts of copra meal on growth of pigs. *Phil. Agric.*, **19,** 111–117.

Hammond, J. (1957). *The growth of the pig.* Pig Progress, Pig Marketing Board: N. Ireland.

Hew, V. F. and Hutagalung, R. I. (1972). The utilization of tapioca root meat (*Manihot utilissima*) in swine feeding. *Malaysian Agr. Res.*, **1,** 124–130.

Klussendorf, R. C. (1957). Stress and animal health. *Cornell Veterinarian*, **43,** 126.

Lim Han Kuo and Ycap Gim Sai (1966). The occurrence of aflatoxin in Malayan imported oil cakes and ground nut kernels. *Malaysian Agr. J.*, **45,** 232–244.

Maner, J. H., Buitrago, J. and Gallo, J. T. (1972). Protein sources for supplementation of fresh cassava. (*Manihot esculenta*) rations for growing-finishing swine. *Jour. Anim. Sci.*, **31,** Abst. 203.

Maust, L. E., Pond, W. G. and Scott, M. L. (1972). Energy value of cassava-rice bran diet with and without supplemental zinc for growing pigs. *Anim. Sci.*, **35,** 953–957.

McNamara, P. (1967). A guide to pig feeding. *Jour. Agr. Western Australia*, Bull. 3517.
Muller, Z., Chon, K. C. and Nah, K. C. (1974). Cassava as a total substitute for cereals in livestock and poultry rations. *World Anim. Rev.*, **12**, 19–24.
Nasol, R. L., Fajardo, R. C. and Rigor, E. M. (1970). Feed-hog production function and maximum profits in Large White hogs. *Anim. Husb. and Agr. Jour.*, **5**, 16–17.
Preston, T. R. (1965). Sugar cane products as energy sources for pigs. *Nature*, **219**, 727–728.
Robertson, G. L., Casida, L. E., Grummer, R. H. and Chapman, A. B. (1951). Some feeding and management factors affecting age at puberty and related phenomena in Chester White and Poland China gilts. *J. Anim. Sci.*, **10**, 841–866.
Savella, H. S. (1963). The African snail is a good hog feed. *Coffee and Cacao*, **6**, 240.
Self, H. L., Grummer, R. H. and Casida, L. E. (1955). The effects of various sequences of full and limited feeding of the reproductive phenomena in Chester White and Poland China gilts. *J. Anim. Sci.*, **12**, 954.
Sewell, R. F. and Carmen, J. L. (1958). Reproductive performance of swine fed chlortetracycline over several generations. *J. Anim. Sci.*, **17**, 752–757.
Sorenson, A. M., Jr., Thomas, L. B. and Gossett, J. W. (1961). A further study on the influence of level of energy intake and season on reproductive performance of gilts. *J. Anim. Sci.*, **20**, 347–349.
Speer, V. C., Brown, H., Quinn, L. and Catron, D. W. (1959). The cessation of antibody absorption in young pigs. *Immunol.*, **83**, 623–634.
Williamson, G. and Payne, W. J. A. (1978). *An introduction to animal husbandry in the tropics*. 3rd ed. Longman: London.

5 Diseases and parasites

One of the major problems confronting the pork producer in the Tropics is the high pig mortality rate due to disease. The continuous presence of destructive protozoan, helminthic and arthropod parasites in humid climates aggravates the difficulties of pig production.

It is reported in the Philippines that the mortality rate due to swine fever, swine plague and other diseases from birth to maturity, is approximately 50 per cent. In addition, liveweight loss due to morbidity, parasitism and crippling afflictions is enormous.

The most effective control measure is prevention rather than cure. Inexpensive control measures for pig diseases that could be easily adopted in the Tropics are as follows:

1 quarantine of premises where the disease occurs;
2 proper disposal of infected and exposed pigs by slaughter and burial or burning; and
3 cleaning and disinfection of premises and equipment.

In more developed temperate countries, control of diseases and parasites can be maintained by the use of disease-free stock. This approach, however, may be uneconomical in developing countries because the management and feeding programmes call for practically sterile conditions.

It is not the purpose of this chapter to provide an exhaustive presentation of pig diseases and parasites but rather to serve as a guide for practical health control practices in pork production. Some details of the principal diseases and internal and external parasites and their control are outlined below.

Specific diseases of the sow

Certain infectious diseases frequently cause abortion, the birth of weak piglets and a high mortality rate among newly born pigs. In some instances, pigs that appear normal die as the result of uterine infection which may cause either loss of milk or the production of toxic agents in the remaining milk. Problems of infertility may also arise from a variety of causes which may be present at the time that the sow is most susceptible to them. The most common infections attacking the genital or reproductive system of the sow are discussed below.

Contagious abortion (Brucellosis)

Brucellosis in pigs is caused by the specific organism *Brucella suis* and results in temporary or permanent sterility, stillborn or weak pigs, abortions, posterior paralysis or lameness and irregular oestrous. In boars, one or both testicles may be inflamed (**orchitis**), and this condition may result in sterility.

Pigs become infected with brucellosis by consuming feed and/or water contaminated with body discharges from animals already infected with the disease. The organism *B. suis* may also enter the body through the reproductive tract during mating.

Abortion, when the sow is 2 or 3 months pregnant, is one of the major symptoms of brucellosis. The agglutination test has been an effective means of diagnosis of the disease. A veterinarian should be consulted regarding this test.

Since no satisfactory treatment or vaccine has yet been formulated for this infection, control and prevention depend strictly on a rigid programme of testing and of disposal of infected animals. If the number of abortions is high, it may be necessary to replace the entire breeding herd. The breeder must then make certain that the animals to be used as replacement stock are purchased from a herd that is certified negative for the disease.

Leptospirosis

Leptospirosis is a disease of sows which is not common in most tropical countries. Abortions may occur in herds infected for the first time, but their incidence diminishes as the herd develops immunity. Most abortions occur 2–4 weeks before the scheduled farrowing. In infected herds, the most common symptom is a large number of weak pigs with high mortality rates.

Wild animals have been found to be carriers of this disease. Most often, the organism that causes leptospirosis is spread by means of the urine from carrier animals.

Vaccination is both economical and efficient in controlling leptospirosis.

Mastitis and agalactia

Mastitis and agalactia are difficult to control. They cause starvation of baby pigs. Mastitis is an inflammation or infection of the mammary gland. Although agalactia, which is influenced by an inherited factor, may occur independently, it frequently results from mastitis; it is manifested by a failure of the sow to produce milk.

The multiple bacterial agents causing mastitis may enter the teats as a consequence of bites from the needle teeth of baby pigs. Dirty and damp farrowing pens

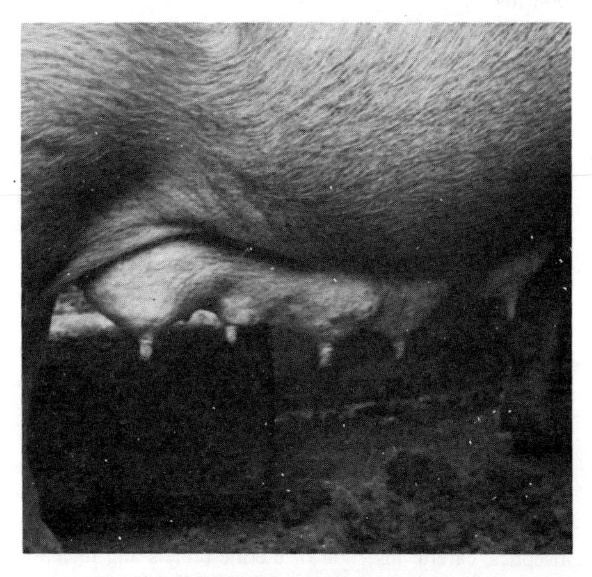

fig. 5.1 Sow with mastitis

provide a favourable environment for the infection of the injured teats and/or mammary glands. When mastitis is at its acute stage, the infected udder becomes swollen, congested, hot and painful. The sow may not allow the nursing pigs to suckle even from healthy mammary glands, if one or two glands are suffering from mastitis (Fig. 5.1).

In chronic mastitis the gland is swollen but firm, cold and painless; there is agalactia or the production of a little milk that is mainly water and sometimes contains clots. Antibiotic infusions into the udder and/or external application of a hot compress and mild antiseptics help to relieve the condition.

In some cases agalactia arises from non-specific or unknown causes. Problems of lactational failure, that is a decrease in the production of milk, may be due to environmental factors such as poor nutrition or high ambient temperatures and genetic make-up such as an inherent hormonal deficiency. When it is due to a deficiency in hormones, a temporary remedy is to encourage lactation by an injection with a pituitary extract containing oxytocin or prolactin. Sows treated in this way should be disposed of after their piglets have been weaned.

Metritis

Metritis is a non-specific infection resulting in an inflammation of the uterus. It arises from several factors of reproduction such as mating, pregnancy, abortion and retention of placenta or dead pigs that cause blood poisoning in the sow. If the sow survives she may become sterile.

Early clinical signs of metritis become evident 2–5 days after farrowing. The animal has fever and chills and discharges from the vulva a sticky whitish-yellow mucus substance that has a foul smell. It may arch its back when walking and show incoordination of movement in its hindquarters. As a rule, sows that have a history of breeding trouble should be removed from the herd.

While metritis should be treated by a veterinarian, there are a few things the pig producer should be able to do. Dead pigs or placenta can be removed by inserting the hand into the genitalia. The hand of the operator must be well disinfected. Flushing of the genitalia with mild antiseptics, careful insertion of uterine tablets and injection of antibiotics all help in arresting the infection.

Other diseases

African swine fever

The causative virus of African swine fever is very resistant, and it has not been possible to immunise animals with an effective serum.

The incubation period, which is the time between infection and the first appearance of symptoms, is 5–15 days. The disease begins with a sudden rise in temperature which lasts approximately 4 days. Unlike the condition in common swine fever, the clinical symptoms appear about 2 days after the temperature has started to decline. The pig eats and moves about normally. Then 2 days before death, it stops eating, becomes depressed, lays in one corner of the pen and refuses to move. General weakness, especially in the hind leg, is evident. Coughing, respiratory disorders and diarrhoea occur. Bleeding is sometimes observed in the edges of the ears, the snout, legs and belly (Fig. 5.2). There is a bloody discharge from the nose and throat, and a marked drop in white blood cell count. Sometimes there are convulsions followed by muscular tremors. Dehydration is pronounced,

Control and prevention The disease is very difficult to control since no appropriate cure or effective serum or vaccine has yet been found. It easily spreads through contamination. The resistant virus can withstand high ambient temperatures and can thrive even in dried or decayed meat. The virus is present in all organs, blood and other secretions of the animal's body.

Prompt diagnosis, adequate quarantine and the subsequent slaughter of infected animals are necessary. The

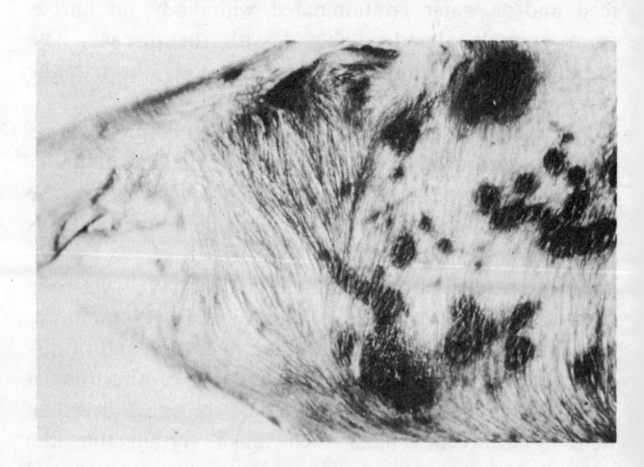

fig. 5.2 Pig with African swine fever
(a) bleeding from the head

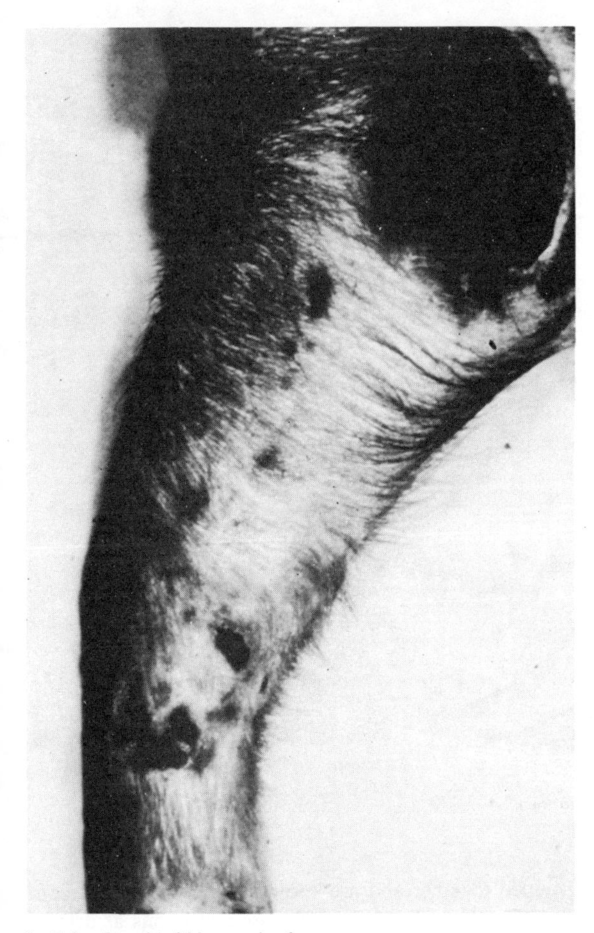

fig. 5.2 Pig with African swine fever
(b) bleeding from the leg

greatest care and strictest hygiene at farms, slaughter houses, harbours and airports in countries where pigs are already infected, can help to control further spread of African swine fever.

In Africa, where the disease is transmitted by wart hogs and bush pigs it is difficult to eradicate it completely. However, it can be controlled by segregating domestic pigs from the wild species.

Hog cholera (swine fever)

Hog cholera is the most serious tropical pig disease and is characterised by high morbidity and mortality rates. All age groups of pigs may be affected and the disease spreads easily, therefore prevention is of the utmost importance.

The virus which causes cholera is transmitted directly or indirectly through carriers such as people and animals. The virus is present in the blood, body tissues, urine, faeces and in the secretions of the eyes and nose of an affected animal. Even a scrap of infected raw pork eaten by the pig or feed contaminated by infected urine can introduce cholera into a healthy herd.

When cholera is introduced in a herd, only one pig may be sick. A period of incubation must elapse before the other pigs show symptoms. This characteristic of the disease shows the importance of immediately separating any sick pig from healthy pigs.

The first symptoms of hog cholera are partial or complete loss of appetite, fever with a temperature of 40–41°C, drooping head, weakness and depression. There may be constipation followed in a few days by diarrhoea. The eyes become inflamed and filled with a gummy whitish discharge causing the eyelids to stick together. Later, brownish crusts may cover the edges of the eyelids. Purple or reddish discoloration of the skin of the belly, ears and legs is frequently observed. Symptoms of pneumonia, like coughing and difficulty in breathing may be manifested. The animal shows incoordination in movement with convulsive attacks (Fig. 5.3). Once an animal shows visible signs of the disease, the possibility of recovery is very doubtful. Death follows in 4–7 days.

An animal which dies of hog cholera may reveal some lesions upon post mortem examination. Small dark red spots indicative of haemorrhages can be observed on the covering of the kidneys, surface of the lungs and outer and inner coverings of the stomach and intestines. Small red blotches are found in the lining of the bladder and the heart.

Prevention and treatment The onset of hog cholera is usually sudden and the spread is very rapid. There is no dependable treatment. Anti-hog cholera serum, an immunising agent with antibodies against hog cholera, has a curative effect when used in the early stages of the disease. Always call a veterinarian at once if hog cholera is suspected.

All herds should be constantly protected against hog cholera by regular immunisation using an effective vaccine. Weanling pigs are best vaccinated 1–2 weeks after weaning. Cleanliness of the pig house, regular disinfection of pens and outdoor lots, and care in feeding help keep the pigs free of disease. If there is an outbreak of hog cholera

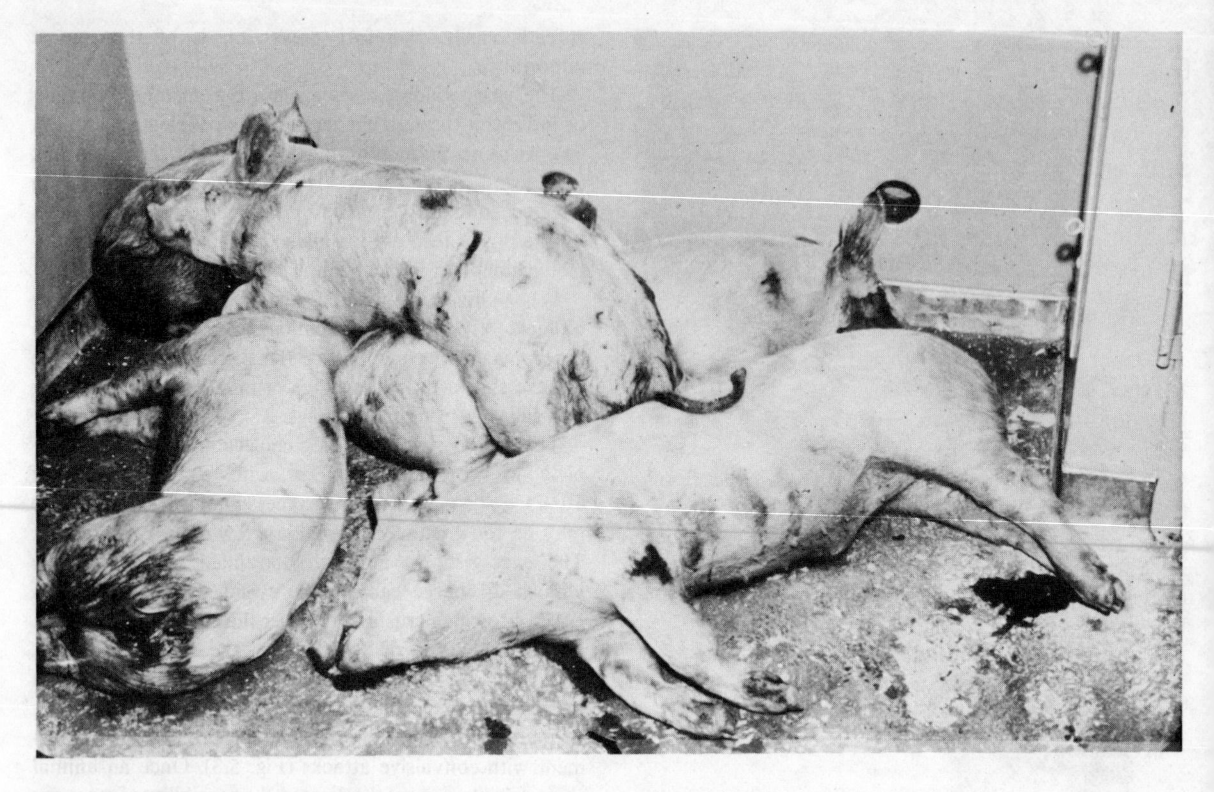

fig. 5.3 Pigs with hog cholera, note bleeding from the nose and bloody diarrhoea

in the region or on a neighbouring farm, great care should be taken in safeguarding the herd. Kitchen left-overs should be fed only to immunised pigs.

Pneumonia complex

There are some respiratory diseases which, although caused by different agents, have similar clinical manifestations. In many instances, contributory factors play an important role in their cause. Poor nutrition, parasite infestation, poor management practices and adverse environmental factors such as high ambient temperatures, high humidity, and inadequate ventilation in housing facilities are all factors that predispose pigs to pneumonia. Pneumonia seldom occurs as a primary infection but more often as a consequence of the presence of several of these stress factors that weaken the resistance of the animal. This group of respiratory diseases can be grouped together and are known as the pneumonia complex. There are four major disease conditions as detailed below.

Swine plague or haemorrhagic septicaemia

It has been reported that pasteurellosis in pigs appears to be a major disease in Southeast Asia, although in most temperate-zone countries the pathogenicity of the organism has apparently decreased. The disease may occur as a pulmonary form that is essentially a type of pneumonia. This form of pneumonia is usually due to the complicating presence of virus pneumonia or swine influenza.

The organism, *Pasturella suiseptica*, may be present in the pig without doing any harm until the vigour of the animal is lowered. Some secondary predisposing and stress factors, such as weaning, shipping, inadequate housing, and improper feeding lower the resistance of the animal and enhance the onset of pasteurellosis in the herd.

Early signs of pasteurellosis are loss of appetite, fever,

chill, exhaustion and a gurgling sound while the animal is breathing. A laboured breathing is quite common in advanced stages of the disease; the pig assumes a sitting position extending its head and breathing through the mouth. When the symptoms become serious, toxaemia and septicaemia may soon set in. A veterinarian should be consulted for advice on the control of the disease.

Prevention and treatment In tropical countries, the environmental stress factors should be minimised by maintaining the health and vigour of the herd by the provision of proper nutrition and housing.

A sick pig should be isolated in clean, comfortable quarters. Proper care and nursing, clean comfortable bedding, fresh water and succulent feeds, such as sweet potato vines and kang kong (*Ipomoea reptans*) are some of the best measures that can be taken to help the sick pig to recover.

Antibiotics, sulfa drugs and nitrofurans, together with analeptics or respiratory and circulatory stimulants (camphor, guaicol, eucalyptin) given in their proper dosages will aid sick pigs to recover from swine plague.

Vaccination with swine plague vaccine must be regularly scheduled, well ahead of the season of high incidence of the disease.

Swine influenza

This disease is caused by the combined action of a filtrable virus and the organism *Haemophilus influenza suis* when stress factors have lowered the vitality of the animal. Young pigs are most susceptible to the infection, particularly when they are badly infected with ascarids or lungworms. Although the morbidity rate is high, mortality is low. More pigs die, however, if cases become complicated by a secondary infection which results in broncho-pneumonia.

Swine influenza is characterised by fever, anorexia, chills and the sudden prostration of a large part of the herd. The breathing of the pig is jerky and is accompanied by occasional sneezing and coughing. The eyes may be red and swollen, and there is a watery nasal discharge sometimes streaked with blood. The animal completely loses its appetite.

This condition may worsen into broncho-pneumonia which can lead to death. In some cases, the pig may recover, but it will take several days to regain its vigour.

Treatment Since no vaccination or specific treatment for swine influenza is available, the usual antibacterial drugs (antibiotics, sulfa compounds) may be given to prevent or even treat complications caused by other diseases or bacteria present. Isolation, strict sanitation, dry warm quarters and fresh drinking water must be provided to hasten the animal's recovery. When sanitation is poor, especially in the garbage-feeding system of pig raising, secondary complications frequently occur, accompanied by severe losses.

Verminous pneumonia (lungworm infection)

Parasitic pneumonia is caused by a whitish thread-like worm (*Metastrongylus* spp.) which is found in the bronchioles, bronchi and trachea of pigs. The irritation and damage this parasite causes in lung tissue predisposes the pig to secondary complications resulting in pneumonia.

Secondary pneumonia

This is a form of pneumonia caused by various bacteria and usually occurs as a secondary complication of a primary condition, such as hog cholera, swine plague, verminous pneumonia or swine influenza. These secondary organisms increase the seriousness of the symptoms and lesions resulting from the primary disease. Secondary pneumonia becomes established only in the lungs of pigs previously affected by some other disease. In many instances pneumonia can be recognised only at the stage when complications have set in. This is why many primary conditions are masked and there seems to be a low incidence of specific primary pneumonia in the herd.

Treatment When pneumonia results from a specific cause, the treatment includes the administration of antibacterial drugs to prevent or possibly cure the secondary complication. A veterinarian should be consulted for proper advice if secondary pneumonia is suspected.

Transmissible gastro-enteritis

This is a sporadic disease of pigs of all ages, and kills many pigs under 10 days old. The losses are less in older pigs. The disease is caused by a filtrable virus.

Among older pigs, especially brood sows, vomiting and profuse diarrhoea are commonly observed. Older pigs recover from the disease promptly. However, a nursing sow, when affected, may stop lactating and eventually lose weight.

In young pigs, diarrhoea and vomiting are always present; the pigs become dehydrated and emaciated and die in 2–7 days. The faeces, which are characterised by a marked odour, are fluid and vary in colour from whitish to yellowish green. Death losses vary from 100 per cent in

1-week old pigs up to 40–60 per cent in 2- to 3-week old pigs. If the pigs survive, they are likely to remain stunted for some time.

Post-mortem examination of pigs that die of gastro-enteritis show that the contents of their intestines are fluid and vary in colour from whitish to yellowish green. If the pig dies early in the course of the disease, the stomach is filled with curdled milk. The kidneys usually show gross evidence of nephrosis and contain urates.

In older pigs the disease runs a longer course. The lining of the stomach and intestines are engorged with blood and frequently have necrotic areas. The outer portion of the kidney is light in colour, while the inner portion is markedly congested.

Treatment No effective specific treatment for transmissible gastro-enteritis has been found. The spread of the disease may be prevented by moving the sick sows into a place far from the farm or the healthy herd. This system of transferring farrowing sows to a different house will break the disease cycle.

In many cases antibiotics only prolong survival without effectively reducing the mortality rate.

A more practical approach is immediately to kill affected young pigs that show emaciation and to dispose of their bodies, thus preventing the spread of the disease in the herd.

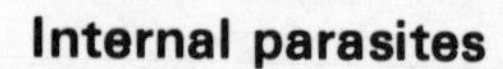

Internal parasites

Ascaris infection (*Ascaris lumbricoides* syn. *A. suum*)

Large intestinal roundworms are common among pigs raised on tropical pastures. Heavy parasitism in pigs is characterised by failure to gain weight, rough haircoat, pot belly, erratic appetite, 'thumps' and weakness. In severe infections, vomiting and diarrhoea, with voiding of live worms may be observed (Fig. 5.4(a)).

Worms may be as long as 30 cm and up to 0·6 cm in diameter. They are normally found in the small intestine but they may also migrate to the stomach and bile ducts when infestation is heavy. The adult female ascarid lays eggs which are voided in the faeces of the pig in the single-cell stage and then embryonate in 10–14 days. Eggs then ingested by the pig from infected pasture hatch in the

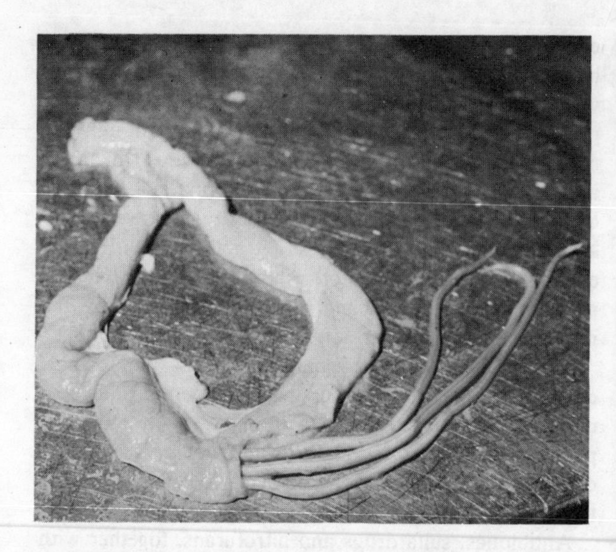

fig. 5.4(a) Roundworms in small intestine of pig

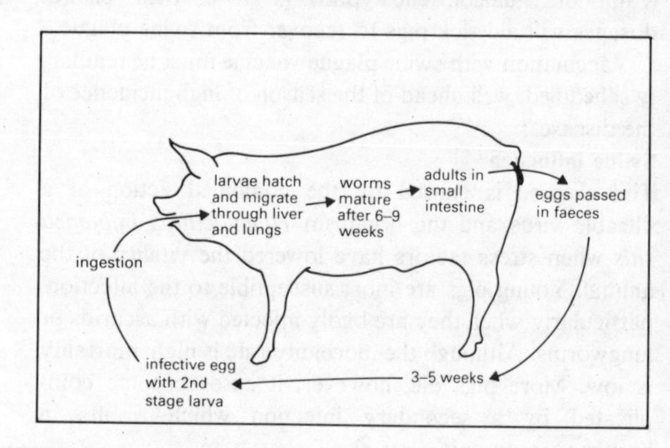

(b) Life cycle of *Ascaris suum*

digestive system and the larvae pass through the walls of the small intestine to the portal blood circulatory system. They migrate into the liver where they cause damage. The larvae may also pass through the heart and migrate to the lungs where the damage that they inflict can produce haemorrhage and cause pneumonia. From the lungs the larvae are swallowed with the saliva and pass into the small intestine where their life cycle is completed and they grow to maturity in 80–90 days (Fig. 5.4(b)).

Prevention and treatment Complete eradication is very difficult to attain. Sanitation and hygiene on the farm are important control measures. Several commercial anthelmintics are available for the treatment of ascarid infected animals. The most popular are piperazine salts which must be regularly administered, especially to growers and mature sows.

Kidney worm (*Stephanurus dentatus*)

Pigs may be infected by adult kidney worms in the renal pelvis, along the ureter and in the peri-renal fat surrounding the kidney. Being erratic and migratory, kidney worm larvae have been located in the brain, spinal cord, lungs, liver and other abdominal organs.

It is estimated that an adult female, 25–45 mm long and 1·8 mm thick, deposits 600 000 to 1 000 000 eggs daily. The eggs, which are partly developed when voided in the urine, hatch in 24–48 hours under favourable conditions. The second-stage larvae emerge in 3–5 days and become highly infective.

Infective larvae are sensitive to dryness and sunlight and therefore are only found in abundance in tropical humid areas beneath trash and litter, in feed and water troughs or in damp places where they can swim.

Pigs can be infected either by ingestion of the larvae or by skin penetration. The larvae are carried by the portal blood circulatory system to the liver where they stay for 2–3 months, causing great damage to that organ. The developing worms in the liver migrate to the kidneys and peri-renal regions and a great number of larvae are lost in the abdominal cavity. Those that settle in the peri-renal tissues form cysts with fistulas. The time required for development from egg to egg-laying adult worm is approximately 6 months, and the symptoms are only seen in pigs over 6 months old (Figs 5.5(a) and (b)).

Symptoms of kidney worm infection are very indefinite, and may be similar to those of other parasitic diseases. Stunting and emaciation, dropsy of the abdomen and posterior paralysis are commonly observed. Posterior paralysis is caused by migration of the worm into the spinal cord or central nervous system.

On post-mortem examination, various abdominal organs, particularly the liver, may show abscesses in different stages of healing. Liver adhesion to adjacent organs may be noted. There may also be a swelling of local lymphatic glands. Pus-filled cysts may be seen in kidneys and ureters. If the kidneys are cut across, adult worms about 2–5 cm long with a thick white and black body may be seen.

Prevention and treatment There is no known medicine for the treatment of kidney worm infection.

fig. 5.5(a) Pig's kidney infected with kidney worm

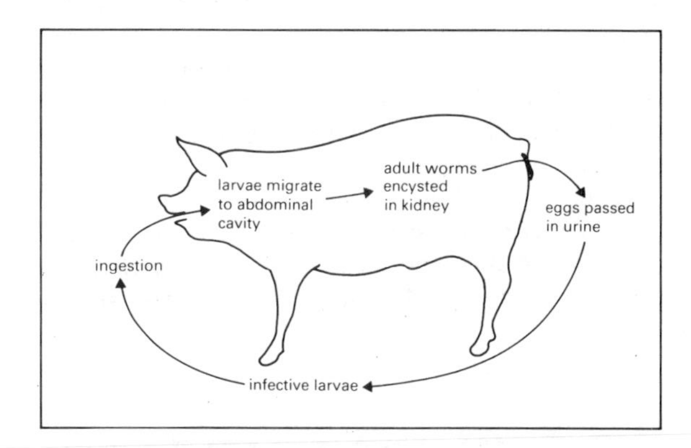

(b) Life cycle of *Stephanurus dentatus*

As kidney worms like damp and moist places, the construction of good drainage ditches is an important preventive measure. The most effective control of kidney worm is the confinement of all pigs on concrete floors in well-kept facilities. Since pigs managed on pasture are most often affected by this disease, maximum sunlight and dryness in the areas surrounding watering and feeding devices should be the aim.

The following managerial practices are recommended in order to minimise losses in areas infested with kidney worms.

1 Use only gilts for breeding and market them when their piglets are weaned.
2 Eliminate all older pigs from the raising area.
3 Practice strict confinement of pigs on concrete floors, especially during their growing period.

The idea of using 'gilts only' for breeding purposes was conceived by scientists at the Beltsville Parasitological Laboratory in the USA. Research suggested that kidney worm parasites may require about 1 year to reach the egg-laying stage so that if only young breeding animals are used for 3 or 4 farrowing seasons or for 2 years, the parasites will be eliminated. In effect, the use of the 'gilt only' in the breeding programme eliminates the kidney worm hosts as the sows are removed and marketed.

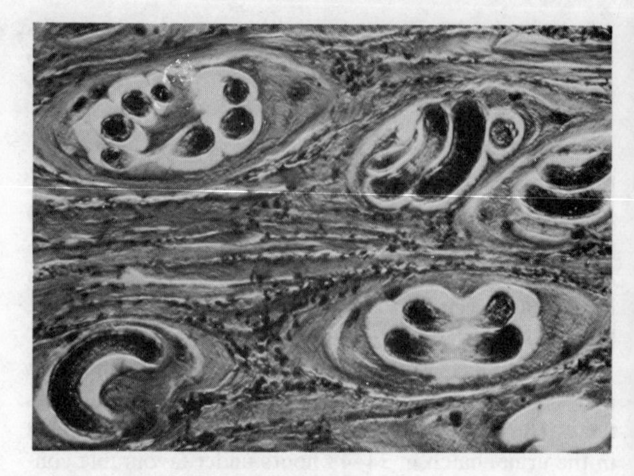

(b) Developing eggs and larvae in pork

Trichinosis

Trichinosis is caused by the swine trichina worm (*Trichinella spiralis*). Although its presence has been reported in several animals including wild mammals, it is primarily found in pigs and men. The worm is usually associated with managerial systems where uncooked food scraps are fed to pigs.

Trichinae are slender thread-like worms present as adults in the small intestines of pigs, as migrating larvae in the blood and as encapsulated worms in the muscles. The adults are from 1·5 to 4·0 mm long and about as wide as a fine thread. The migrating larvae are microscopic in size and the encapsulated larvae are spirally rolled and about 1·0 mm long (Figs 5.6(a), (b) and (c)).

Transmission and control Proper management of pigs is very important in preventing the spread of this parasite. The best method is not to feed the pigs with uncooked kitchen left-overs and food scraps. Educating the public concerning the importance of properly cooked pork is also important in preventing the spread of the disease to man.

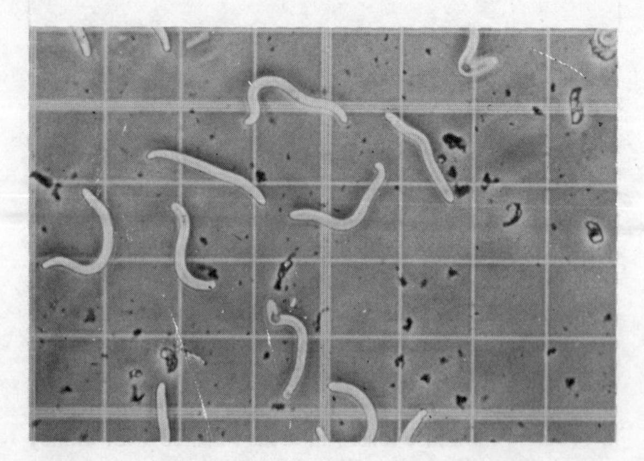

fig. 5.6(a) *Trichinella* larvae in pig blood on grid for counting

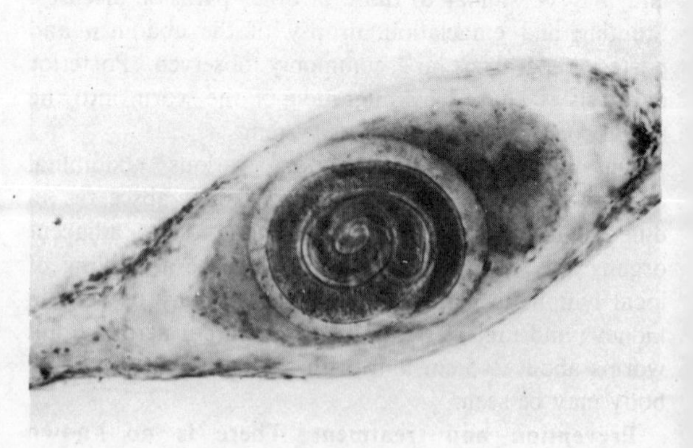

(c) Encapsulated larva in pig diaphragm

fig. 5.7 Pig with dry, scaly hide caused by mange

External parasites

Hog mange and lice

Mange of pigs is a highly contagious condition. It stunts the growth of the infested pigs and can cause death. It is caused by an almost microscopic mite which spends its entire life cycle on the pigs.

Mange causes intense itching and mangy pigs continuously rub themselves against hard objects. The mites dig themselves into the skin of the animal causing a characteristic dry, scaly hide (Fig. 5.7). They feed on the tissues and blood of the animal, appearing first around the eyes, nose and ears, then spreading all over the body. Mange mites favour infested quarters, especially moist bedding.

Pig lice are the largest of the blood-sucking lice which attack livestock in the Tropics. The louse is about 6·4 mm long and greyish brown in colour (Fig. 5.8). Lice reduce liveweight gains by their blood-sucking habits as well as by the irritation they cause. They are commonly found in small groups around and in the ears, in the folds of the skin around the neck, inner upper lips and the tail. The females glue their eggs to hairs; the eggs are laid at the rate of 3–6 eggs a day per female. The young lice emerge from the eggs in 2–3 weeks and become fully grown in another 2 weeks.

Control All control measures against hog mange and lice should include thorough cleaning of shelters and houses and the removal of all manure followed by disinfection with chemicals, steam or boiling water. Floors and walls of pig houses should be sprayed with insecticide. At the same time that the animals are treated, their bedding should also be cleaned.

Insecticides, such as lindane and benzene hexachloride are excellent chemicals to use in the control of mange and lice. A 0·1–0·25 per cent solution of either has been

fig. 5.8 Pig louse (*Haematopinus suis*) magnified × 16

found effective. Used as a spray 2·0 to 5·0 litres of solution is sufficient for the treatment of 1 animal. Pigs may also be dipped using a similar solution.

Treatment of the sow before farrowing will keep her litter free from mange. Sows should be treated at least 30 days before farrowing. When it becomes necessary to treat sows with a litter, special care in the mixing and application of the chemicals must be taken. The sow's teats should be dried before allowing baby pigs to suckle.

In countries where the use of DDT is not prohibited pigs infested only with lice can be sprayed with 2·0–5·0 litres of a 0·5 per cent DDT solution, or they may be dipped in a similar solution.

Prevention of disease

All efforts must be directed towards keeping the herd's health at its best. Only good healthy pigs can develop sufficient resistance to diseases. The vigour of a herd can be enhanced by proper nutrition and feeding.

The causative agent of virus pneumonia must not be allowed a chance to attack the animals. When a farm, however, has been affected by virus pneumonia for a long time, it must be assumed that the sows are also carriers. This is especially the case with young sows. It is recommended that under these circumstances young breeding sows be kept for only a period of 18 months after they have reached breeding age (maturity), and that older sows be farrowed in isolation. This will result in virus-free baby pigs from the litter crop of virus-free younger sows.

Further reading

Eusebio, J. A. (1969). *The science and practice of swine production: with emphasis on Philippine conditions.* UPCA Textbook Board: Los Baños, Philippines.

Hall, H. T. B. (1977). *Diseases and parasites of livestock in the Tropics.* Longman: London.

Richards, R. B. (1969). Practical disease prevention in the pig herd. *J. Agric. Western Australia*, **4,** 9.

United States Department of Agriculture. (1952). *Diseases of swine.* Farmers' Bulletin No. 1914. USDA: Washington.

Young, G. A. (1958). *Control and elimination of swine diseases through repopulation with disease-free stock. Diseases of swine.* Iowa State University Press: Ames, Iowa.

Part 3 Pig husbandry

6 Care and management

Modern pig management includes feeding, housing, care and control of diseases and other well-planned practices that are essential for increasing efficiency in pig production. In most developing tropical countries, especially in remote rural areas, pigs are raised as scavengers. The least effort at proper management practices is expended to secure maximum productivity.

Some of the management practices recommended in this chapter are already in use in some tropical countries. Housing and equipment are considered in the next chapter.

Sows and boars

Every afternoon when the temperature is high in the hot months, water should be sprinkled on the boars and pregnant sows to help them reduce their body temperature. This practice will improve the conception rate to a considerable extent. It will also create a favourable environment for the development of the fertilised eggs. Table 6.1 presents data that demonstrates the benefits that can be derived from sprinkling pregnant sows. The experiment was conducted at Fort Remo during June, July and August. The average daily maximum temperature during the 92-day test was 34°C. The average rectal temperature for the non-sprinkled sows was 39·8°C and for the sprinkled sows, 38·3°C.

It has been reported in the Philippines that the largest average litter size was 10·3 for pigs born during the month of April, and 8·5 for pigs farrowed during the month of July. The sows had been fertilised in January and April, respectively, that is, during the coolest month and the hottest dry month of the year. In summer the pregnant sows should be provided with a bath 4 times a day. In Colombia concrete wallows are provided for sows that are already pregnant (Fig. 6.1 and cf. p. 80).

Do not let the sows grow too fat. Fat sows will not produce as good and as healthy a litter as lean sows.

Table 6.1 Cooling effects of sprinkling water on sows during pregnancy. (*Source:* **Whatley, J. A.** *et al.* (1957). *Temperature studies on sows during gestation.* 31st Annual Livestock Feeder's Day Report. Misc. Publ. No. MP48. Oklahoma A. and M. Agric. Exp. Sta.)

Treatment	With sprinkler	No sprinkler
Sows used	17·00	17·00
Pigs farrowed	10·88	9·24
Live pigs	10·06	7·71
Still-born pigs	0·82	1·53
Decomposed embryos	0·06	0·65
Total litter birth weight (kg)	12·70	11·60
Live litter birth weight (kg)	12·00	9·70
Weaned pigs	7·76	5·71
Litter 56-day weight (kg)	139·38	100·61

However, sows that have been served should be fed a high-energy ration for 1 week. After this period the sows should be placed on a restricted energy intake by providing them with plenty of soilage mixed with their concentrate feed. Alternatively, they may be fed on pasture with concentrate feed. The concentrate should be fed twice daily – once in the morning and again in the afternoon at the rate of 1·5 kg per feed for 14 weeks. If there is a tropical depression or typhoon, however, it is advisable to return the pregnant sows that are on pasture to their pens.

One week before the scheduled farrowing, return the sows to a high-energy feeding programme (2 feeds a day) and keep them in close confinement. Maintain the high-energy feeding programme for the sow's entire nursing period. Feed sows kept in close confinement with soilage, such as kang kong (*Ipomoea reptans*), sweet potato vine and leaves, ramie (*Boehmeria nivea*) leaves, and young maize stalks.

fig. 6.1 Brazilian pigs in a wallow

Overcrowding in the pen for sows that are in an advanced stage of pregnancy usually causes abortion. Each pregnant sow should be allotted a space of at least 4–5 m^2. A boar needs a floor space of 4–6 m^2. Abortion may also be caused by vaccination, thus pregnant sows should never be vaccinated.

Baby pigs

At night or on cold days, heat the pen for baby pigs with a 50 watt electric light bulb. The bulb should be suspended 30–40 cm above the bedding. In the absence of electricity, cover the floor of the farrowing pen with sawdust or some other suitable bedding (Fig. 6.2).

Clip the tips of the needle teeth of baby pigs at birth or on the first day after birth. Use a tooth nipper or sharp-edged pliers. Clipping the teeth will prevent injury to the sow's teats while the pigs are nursing (Fig. 6.3).

Baby pigs in close confinement suffer from nutritional anaemia because they depend entirely on sow's milk, which has a low iron content. This is why the condition known as 'thumps' is common among pigs 2–3 weeks old. Thumps is characterised by difficulty in breathing. To prevent nutritional anaemia in baby pigs, day-old pigs should be injected with 2 ml of iron-dextran, a commercial preparation. This will also improve their growth rate. Injection and dosage should be carried out according to the prescription for the product (Fig. 6.4).

A practical and cheap way to prevent nutritional anaemia in baby pigs is to put in the creep pen, clean soil that is rich in minerals. Alternatively, the sow's udder can be swabbed daily during the nursing period with an iron(II) solution. The solution is prepared by dissolving 0·5 kg of iron(II) sulphate(VI) and 5 tablespoonsful of sugar in 0·94 litre of water. Neither method is as satisfactory as iron-dextran injection.

Scours or diarrhoea is also common among baby pigs 1–3 weeks old.

fig. 6.2 Baby pigs warmed by a lamp

fig. 6.3 Clipping the needle teeth of a baby pig

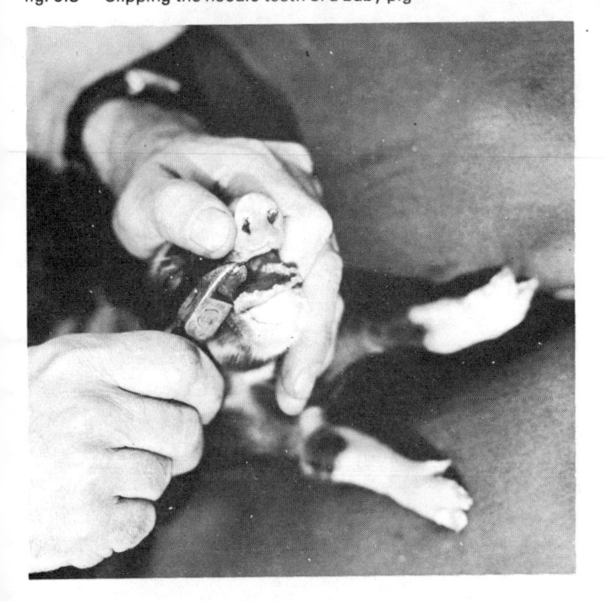

fig. 6.4 Injecting a baby pig with an iron preparation to prevent anaemia

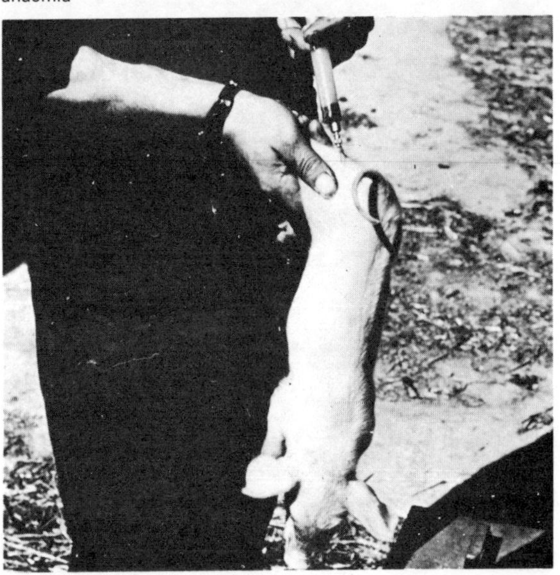

fig. 6.5 Pregnant gilts resting in partitioned pens

fig. 6.6 Sow in farrowing pen with litter

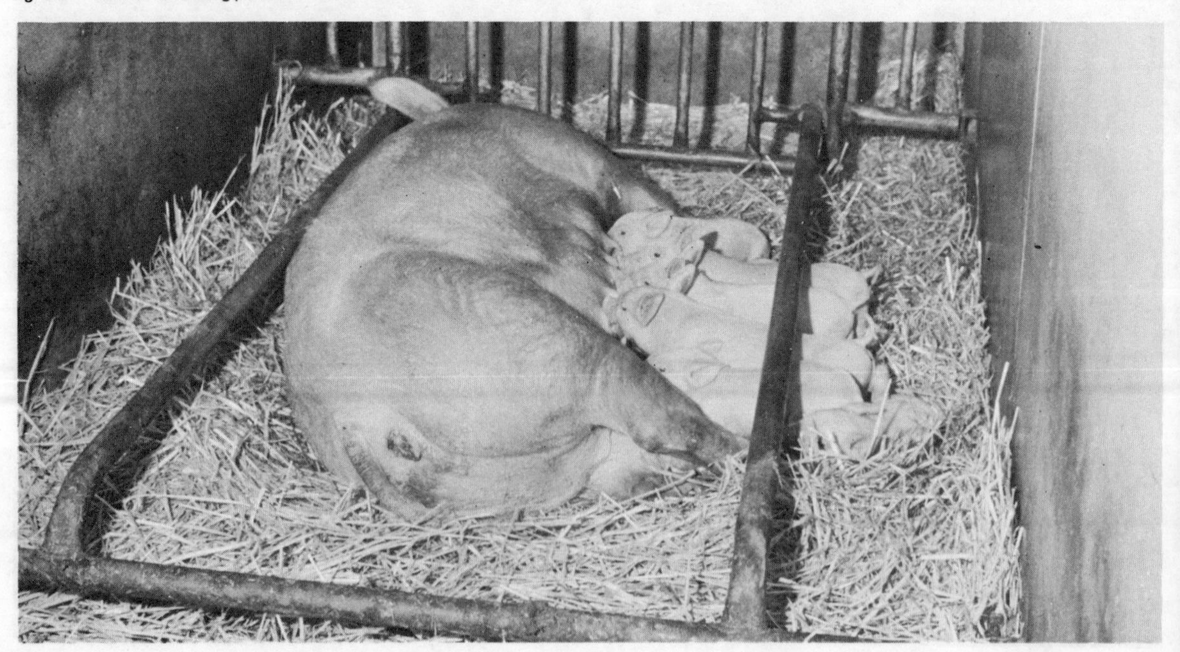

The sow will usually farrow within 24 hours after milk appears in her teats. To keep the litter from being crushed by the mother sow, place her in a farrowing stall 3 days before scheduled parturition. If the pen is shared with other sows, the piglets are likely to be crushed (Figs 6.5 and 6.6).

Good sanitation for the litter should start at farrowing time. Clean the sow and the farrowing stall every day for 3 consecutive days before she farrows. Wash the sow's udder with a 10 per cent creoline solution or with a soap solution before she is placed in the farrowing stall. This will remove any dirt and/or parasite eggs that may be adhering to her body and which may be passed on to the newly born pigs.

If a sow does not farrow more than 6 pigs at a time and is a poor mother, fatten her up after her litter has been weaned, and dispose of her, either by sale or by slaughter. There are some pig producers who do not cull a sow on the basis of the 1st litter performance. They allow a 2nd farrowing without regard for the number and the weight of the pigs in the 1st litter. This is because in some countries in the Tropics, it requires a longer time and a higher cost of feed to grow replacement gilts than to allow the sow to farrow again. On the other hand, a sow that has proved to be prolific and a good mother should be kept as long as she continues to do well. However, after 6 farrowings it is reasonable to sell her. There is strong evidence to suggest that with the average sow, the number of pigs farrowed per litter declines after the 6th farrowing.

The milk secretion of the sow is generally high during the 2nd and 3rd weeks after parturition. The milk produced at this time is rich in fat. The total fat intake by the piglet during this period is therefore often more than its digestive system can tolerate. Also during the early stage of the pig's life, it does not produce a sufficient quantity of the enzyme needed for digesting fat. These two factors reinforce each other and may cause indigestion followed by scours.

Antibiotics will help minimise scouring caused by micro-organism disturbances rather than physiological factors. Add the antibiotics to the piglet's drinking water at the maximum level recommended by the manufacturer. If this does not stop scours, regulate the feed intake of the litter or separate them from the sow at night. Another method is to start the pigs eating the pre-starter ration at 2 or 3 weeks of age. Pre-starter feeding, however, should be done only when the ration is not too expensive. Dried skim milk, the major feed ingredient in the pre-starter diet for baby pigs, is generally expensive in the Tropics.

Many pig raisers practice early weaning of baby pigs at 1 month of age in order to obtain 5 litters in 2 years from their sows. Early weaning may become an economic necessity in large-scale commercial piggeries in countries where pig production is limited, and pork is expensive. In some countries in the Tropics, early weaning may be practical and also economical if the cost of milk by-products is relatively cheap. Early weaning is most successful when it is combined with high-level feeding, rigid disease control and good management. Practical pre-starter diets have been formulated in many tropical countries, such as Colombia, Mexico, the Philippines, Singapore, Thailand, and Taiwan, as a substitute for sow's milk in the diet of early-weaned pigs. They can produce just as good or even better growth rates than those of pigs conventionally weaned at 8 weeks of age. Fig. 6.7 shows a crate used for weighing young pigs.

fig. 6.7 A special crate for weighing young pigs

Whenever possible, early-weaned pigs should be housed in a separate building. This building should be far from the central pig house in order to create a disease barrier. The central pig house and its surroundings are likely to be heavily contaminated.

Undesirable male pigs and those intended for meat should be castrated within 1 week of birth. Early castration is more convenient and in addition, the wound heals faster.

Growing/finishing pigs

Confinement feeding is a developing practice in the feeding management of growing/finishing pigs. It has become more popular because of the availability of balanced rations. Growing/finishing pigs managed in the pig house save approximately 10–15 kg of feed per 50 kg weight gain. Confinement feeding also releases pasture land for other uses, especially in countries where land is urgently required for grain and other food crop production. In addition, it minimises infection from parasites and diseases because the pigs do not come in contact with the soil (Fig. 6.8).

fig. 6.8 These young pigs in confinement feeding are being raised for bacon

The pigs should be provided with a concrete floor space where they can exercise. A floor space of 0·81–1·08 m^2 per growing pig will serve the purpose (see Chapter 7).

When the weather is hot, hose or sprinkle the pigs with water. This practice improves the pig's growth. Growing pigs are very sensitive to temperature changes above 30°C. Investigations have shown that for each degree rise in average daily temperature above 30°C, the daily liveweight gain is reduced, and that the feed required per unit weight gain is increased.

When the pigs reach the profitable market weight of 80 kg they should be sold or butchered. In tropical South American countries marketable weight is 90 to 100 kg. In some other tropical countries it is as low as 60 kg. Heavier pigs can be profitable if the cost of feed is cheap, or if the feed is mainly kitchen left-overs and is not purchased. Dispose of slow-growing animals or those sows and gilts that do not reproduce regularly in spite of good feeding and management.

Castration

Castration of male pigs is a routine practice. Pigs are castrated to prevent the perpetuation of genetically undesirable animals and to improve the meat quality of male pigs for the market. Also, old boars that are no longer useful for breeding may be castrated before they are slaughtered for meat. This removes the male odour flavour inherent in the meat of old uncastrated boars.

Age at castration

All male pigs not intended for breeding should be castrated while young, preferably when they are 1–3 weeks old. The younger the animals the faster they will recover from castration because their wounds are often small and heal faster; in addition, it is much easier to restrain younger animals than the older ones. When the animals are castrated young their growth is not arrested.

Procedure in castration

Do not feed the grower or inactive boar, for one day immediately before the operation. This prevents excessive bleeding and reduces strain on the abdomen while the pig

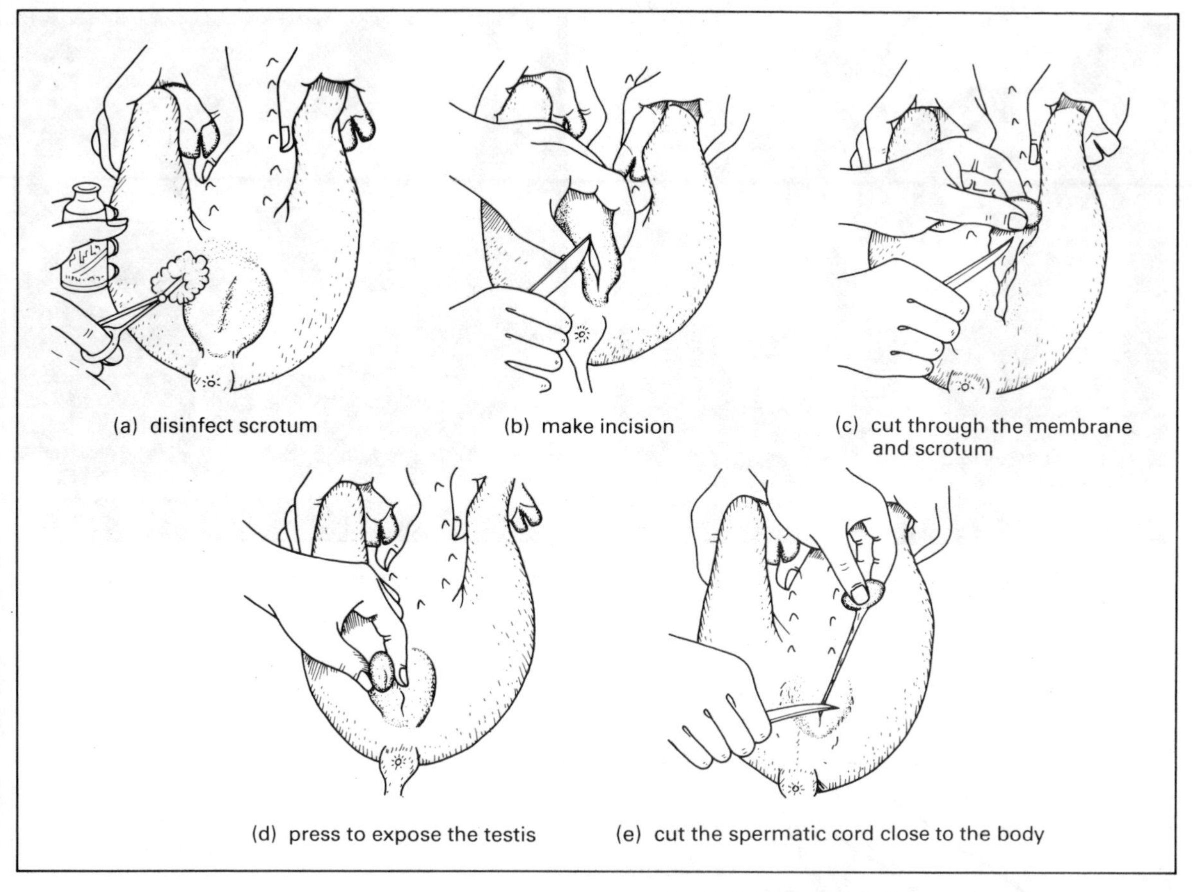

fig. 6.9 Castration procedure

is being restrained. Suckling pigs need not be fasted before castration. Fig. 6.9 illustrates the procedures involved in castration and Figs 6.10(a), (b) and (c) show a young pig being castrated.

Strict hygiene is essential for successful castration. Disinfect all instruments to be used in the operation. The operators should wash their hands with soap and water and rinse with a disinfectant.

Place the pig on a castration rack to make restraint easier (Fig. 6.11). Hold the pig with its back on the rack and its tail facing the person doing the castration. If no castration rack is available, hold the pig by its hind legs with its back away from the helper. The helper clamps his knees on the pig's ribs near the shoulders.

Disinfect the scrotum and surrounding area with a

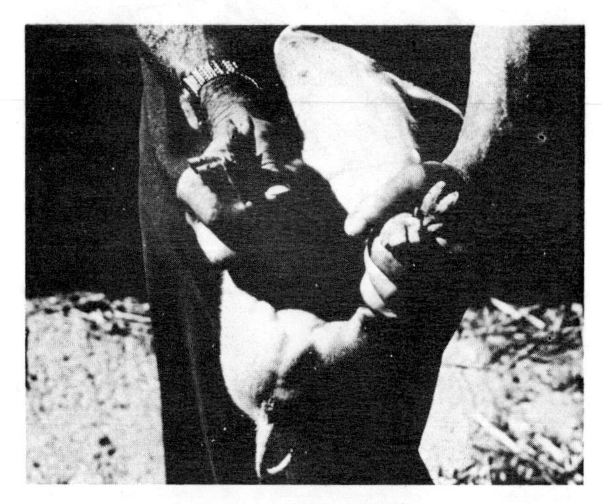

fig. 6.10(a) The pig is held ready for castration

fig. 6.10 (*continued*)
(b) The scrotal sac is cut with a scalpel

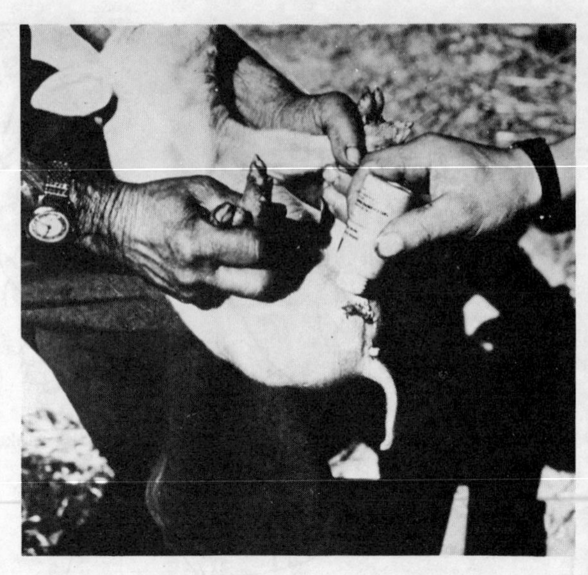

(c) Antibiotic powder is applied to the wound after removal of the testicles

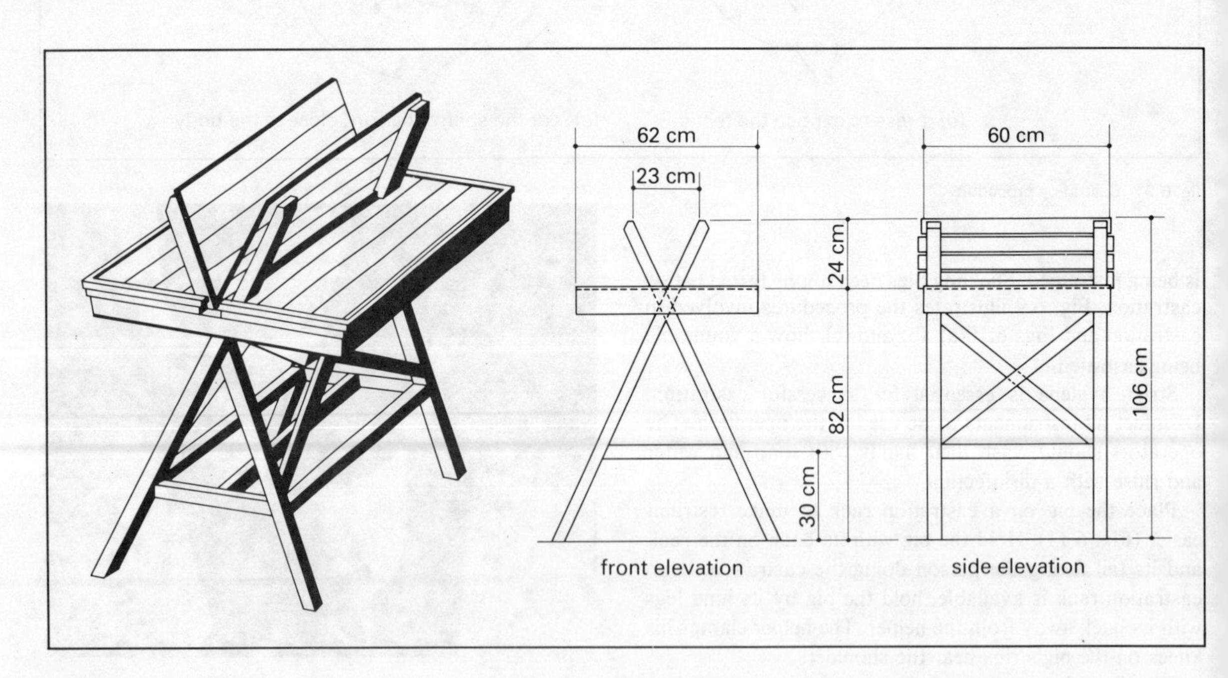

fig. 6.11 Castration rack

potassium manganate(VII) (permanganate) solution (1: 1000 dilution) or tincture of Zephiran.

Using the thumb, forefinger and middle finger, press the scrotum tightly in order to stretch the skin.

Slit the scrotum with a sharp knife. One slit may be made (1-slit method) directly between the two testicles. Or 2 slits may be cut – one on each side of the scrotum (2-slit method). Make the cut as low as possible so that the blood will drain properly. Make the slit deep enough to cut through the scrotum and membrane. If the knife is sharp, one cut towards the operator is sufficient to expose the testicles.

To force the testicles out, press with the fingers behind them and then pull them out. Press the spermatic cord with a forceps to prevent excessive bleeding. If you do not have a forceps, tie the spermatic cord with thread.

Cut the spermatic cord between the testicle and the forceps. Then apply tincture of iodine to the cut. Do the same with the other testicle. With young animals, suturing the wound is not necessary.

Care after castration

After the operation, clean the area surrounding the wound with cotton wool that has been soaked in a disinfectant. Apply iodine to the wound to prevent infection. To keep

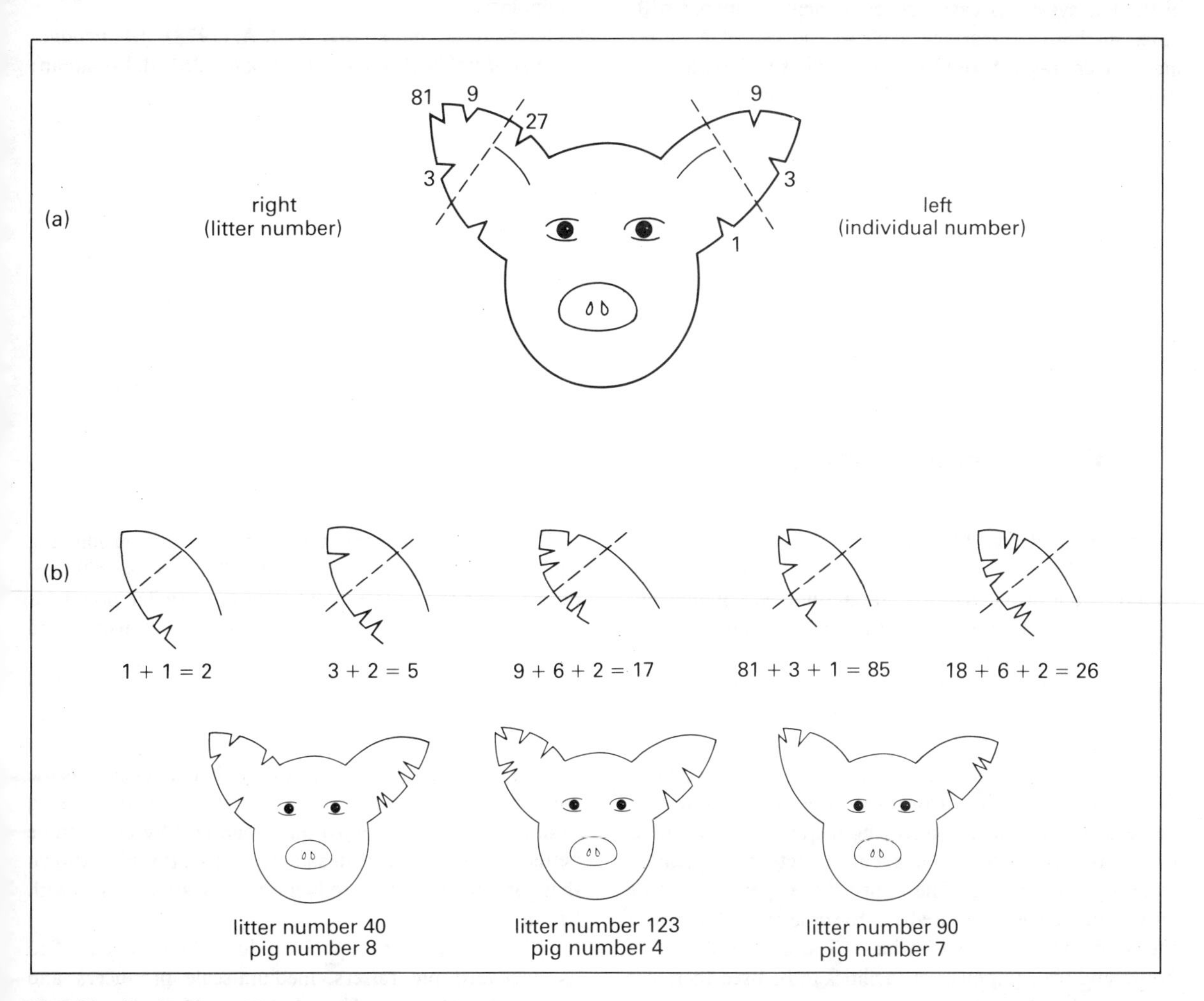

fig. 6.12 Key for ear notching

flies away from the wound, apply tar around it. Keep the animal in a clean, dry pen. It is not necessary to give the animal special nursing care. If an old boar is castrated, feed only one meal ration the day after the operation, when his appetite is restored.

Ear notching

This is an important management practice necessary for the identification of animals for record purposes and it should be carried out at as early an age as possible. Usually ear notching is done not more than one day after birth. The system is easy and permanent compared with other methods of marking such as the use of a tattoo mark or ear tags. A key for ear notching is illustrated in Fig. 6.12(a). The right ear is used for the litter number and left for the individual number. Some examples of ear notching for individual identification are illustrated in Fig. 6.12(b).

Further reading

Arganosa, V. G., Puyaoan, R. B. and de Ramos, M. B. (1970). The performance of pigs born in different years and months of the year. *Phil. J. Anim. Sci.*, **7**, 29.

Eusebio, J. A. (1969). *Care and management of swine.* UPCA Cir. No. 8. Univ. Philippines: Los Baños, Philippines.

Williamson, G. and Payne, W. J. A. (1978). *An introduction to animal husbandry in the tropics.* 3rd ed. Longman: London.

7 Housing and equipment

Confinement of pigs has now generally replaced the old practice of raising them on pasture. In some tropical countries use of mechanised and automated confinement systems is on the increase, although most pigs are still managed under rather primitive conditions.

The advantages of the confinement system are improved feed efficiency, lower production costs, and more efficient control of disease and parasites. Pig raisers claim that the most outstanding benefits derived from confinement are protection from parasite infestation and decreased labour spent on feeding the animals, as well as time saved on other management activities such as watering, breeding, etc. The ability of the operator and the system of management used will to some extent determine the profitability of a pig business. The choice of type of house and the equipment or facilities to be used therefore merit careful attention.

There is no standard type or system of housing. Pig raisers in different countries in the Tropics use different designs, but they have adopted certain similar principles and practices related to the choice of housing and equipment.

In deciding what kind of pig house to construct, pig producers should consider a house that reduces labour input but increases efficiency in management and operation. Another important feature that they should consider is the ease and degree to which good sanitation can be achieved in the house. It is accepted that clean healthy pigs can be better produced in a clean sanitary house. It is also known that profitable pork production is possible only with healthy pigs.

In tropical countries, pork producers may be classified as backyard pig raisers, medium-scale producers and large-scale producers. The other group of producers, very

fig. 7.1 (a) A Nigerian piggery

(b) A Brazilian piggery

fig. 7.1 (*continued*)
(c) A backyard pig house in the Sudan

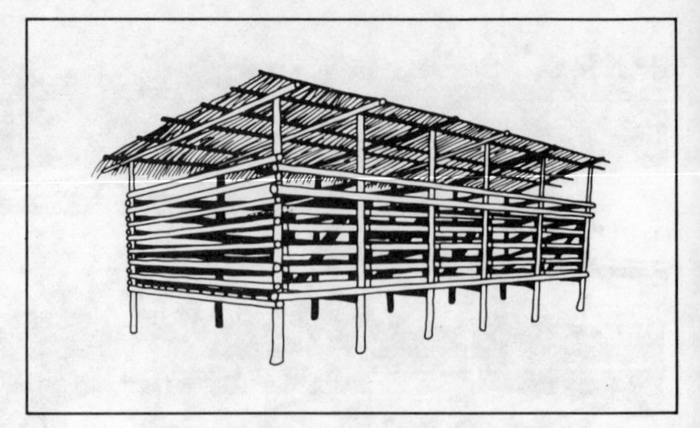

fig. 7.2 Elevated pigsty built from indigenous materials

common in tropical countries, are those who let their pigs wander and scavenge with no attempt to secure maximum productivity from the pigs.

Each type of pig raiser adopts a housing system and facilities appropriate to the size and extent of his pig operations (Figs 7.1(a), (b) and (c)).

Types of housing

Elevated pigsty

As the name suggests, backyard pig raising involves only a few animals (perhaps 1 to 3) which the family raise in the backyard as a kind of 'piggy bank'. A very common countryside scene in many tropical countries is a pig with a rope around its neck tied to a tree alongside the road or to a post under the house. Pigs raised in this way reach market weight after a long period of feeding, compared to those kept in concrete pens. If the pigs are not tied with a rope to a tree or to one of the posts of the house, they may be housed in an elevated battery or pigsty (Fig. 7.2). This type of house may have as many compartments as there are pigs raised by the family. Each compartment should be at least 0·9 × 1·0 m in area. It can be roofed with a grass such as *Imperata cylindrica* or palm leaves such as nipa or coconut, that will help to keep it cool. The house should be made from cheap materials that are easily available in the region. The floor should be slotted and made of whole bamboo or wood. If this is raised some height from the ground, the animals do not come into contact with the soil, and they are less susceptible to infection with parasites.

Generally, raising pigs in this type of housing is cheap and easy. However, it also has a few disadvantages. The slotted floor is not ideal for a pregnant sow as she may slip through a slot and suffer an abortion. The compartments are usually small, so that sows do not have enough space on which to walk for exercise.

Conventional housing

The conventional housing system may combine several managerial operations into a single-unit house (Fig. 7.3) or it may be made up of two units; a central pig house for both pregnant and non-pregnant sows, farrowing and/or lactating sows and growing pigs, and another house in which baby pigs are reared and weaned. This type of housing is adapted to the needs of the medium-scale pork producer who keeps from 5 to 20 sows. Equipment for this type of house is illustrated in Figs 7.4 and 7.5. The information in the illustrations should be taken as a guide and not as an absolute standard for housing.

It is wise to adapt housing construction to local conditions. For example roofing materials should be chosen so that the inside of the house is kept cool during the hot months. It is probable that the most suitable roofing material is thatch, made from a local grass, reed or palm leaf. In places where typhoons or hurricanes are frequent it is best to build the roof a little lower than the height indicated in the illustration. It is most important that adequate water facilities should be provided in all the

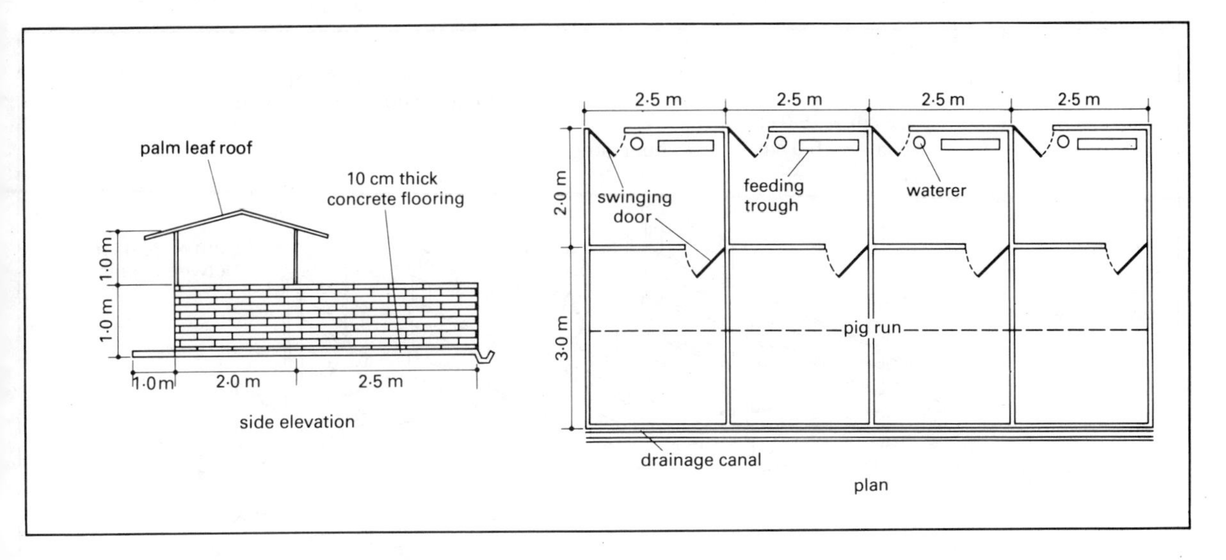

fig. 7.3 Conventional pig houses for the Tropics

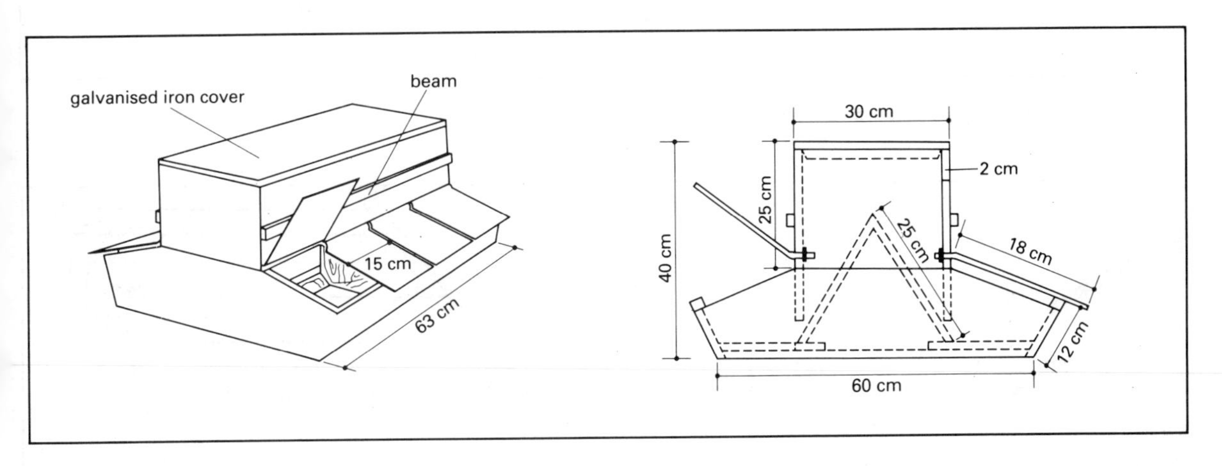

fig. 7.4 Two-sided self-feeder for early-weaned pigs

pens. These should include water for drinking and for keeping the pigs cool. Frequently, central pig houses are provided with a hog wallow that is located at one end (Fig. 7.6). Here the pigs are allowed to bathe during hot weather. However, a hog wallow is difficult to maintain; it has to be cleaned out and the water has to be replaced frequently with clean water, otherwise the pigs may acquire diseases and parasites from contaminated water in the wallow. Overhead sprinkling or bathing the pigs 3 to 4 times daily may be a better way of cooling them (Fig. 7.7).

If the house is intended to be permanent, its floor should be made of concrete. A concrete floor is easier to clean and it can be kept clean more easily than any other floor. Therefore disease and parasite infestations are minimised.

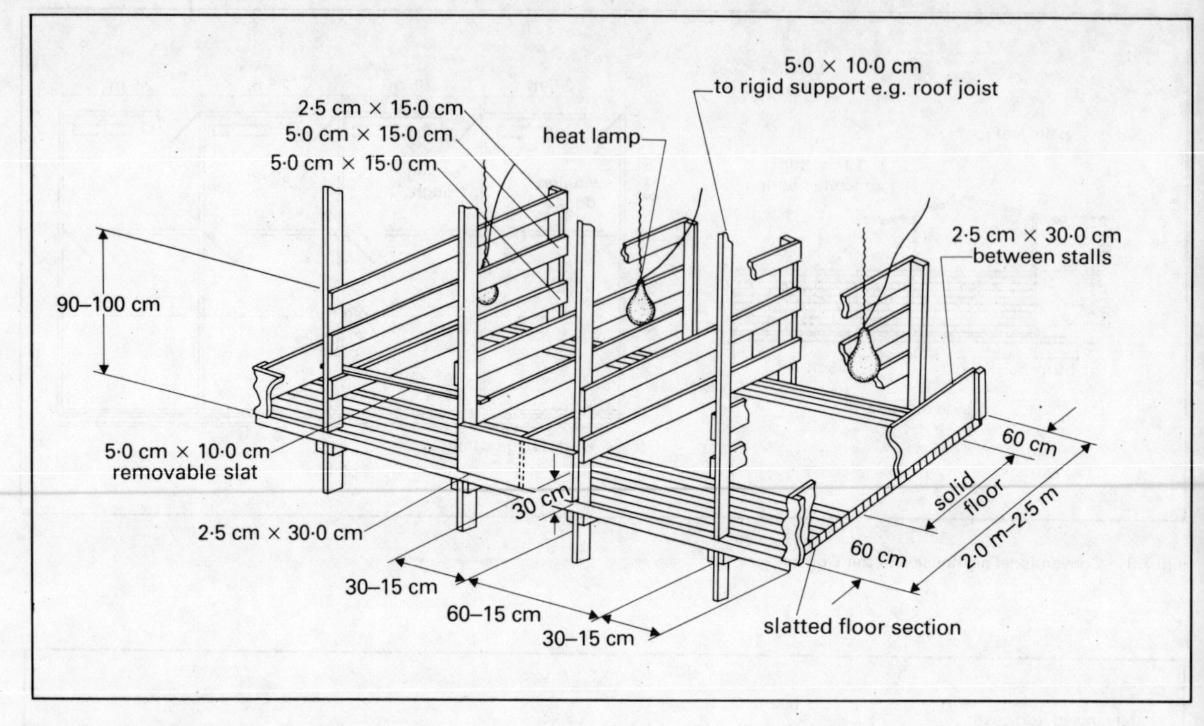

fig. 7.5 Elevated farrowing stall with a slatted base

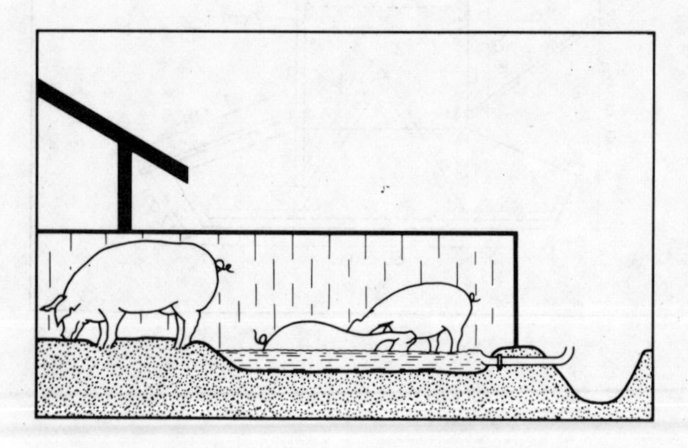

fig. 7.6 Wallow for pigs managed under tropical humid climatic conditions

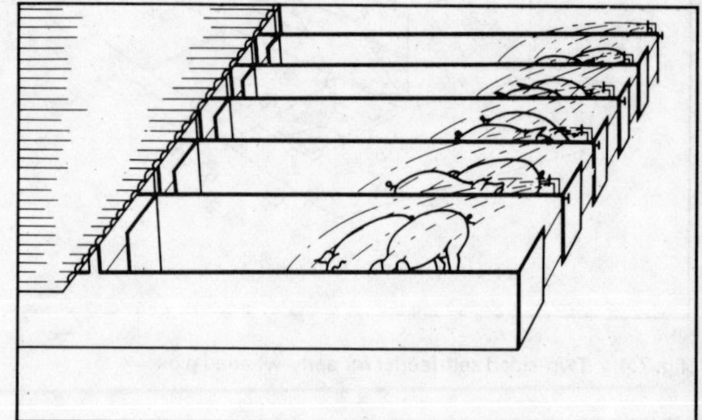

fig. 7.7 Water sprayers installed in pig pens

Life-cycle housing

The newest concept in pig housing and management is life-cycle housing. This system is designed to provide pigs with proper space and comfort during each phase of their life cycle (Figs 7.8, 7.9 and 7.10). In this system, pigs that have the same feeding and management requirements are placed together. This reduces overall space requirements, provides maximum labour efficiency and controls disease. Where 100 or more pigs are kept, the life-cycle housing system is more economical than the conventional system.

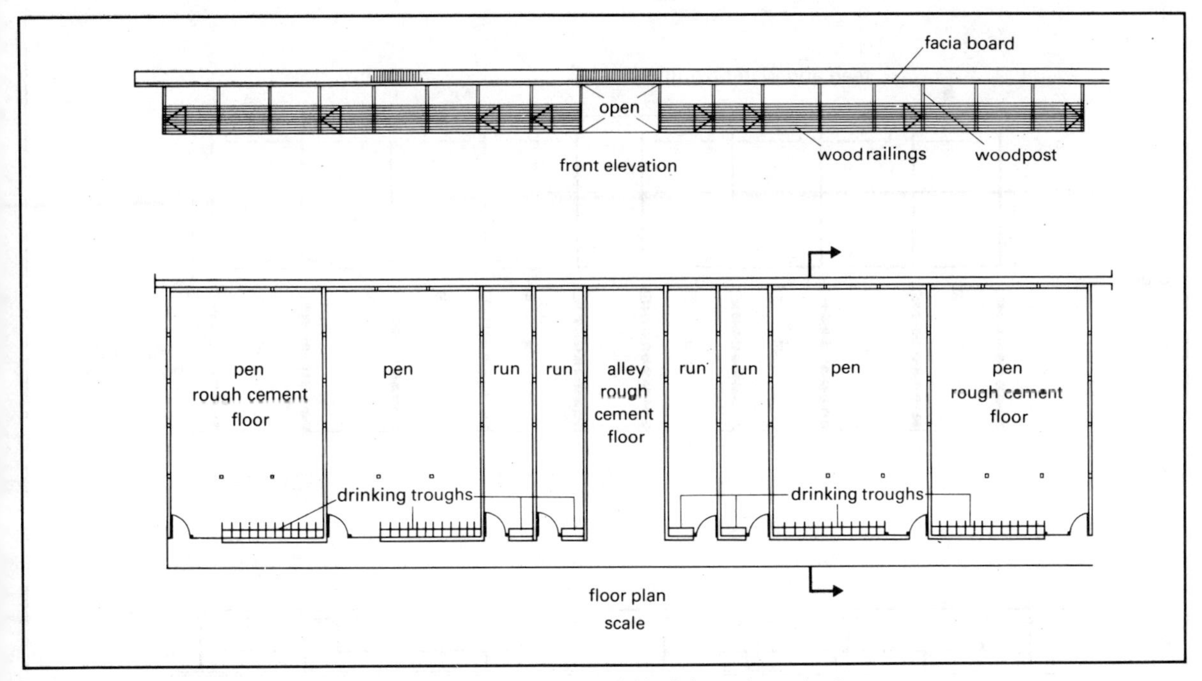

fig. 7.8 Plans for a pre-gestation and gestation house. Boar pens are individually located near the alley

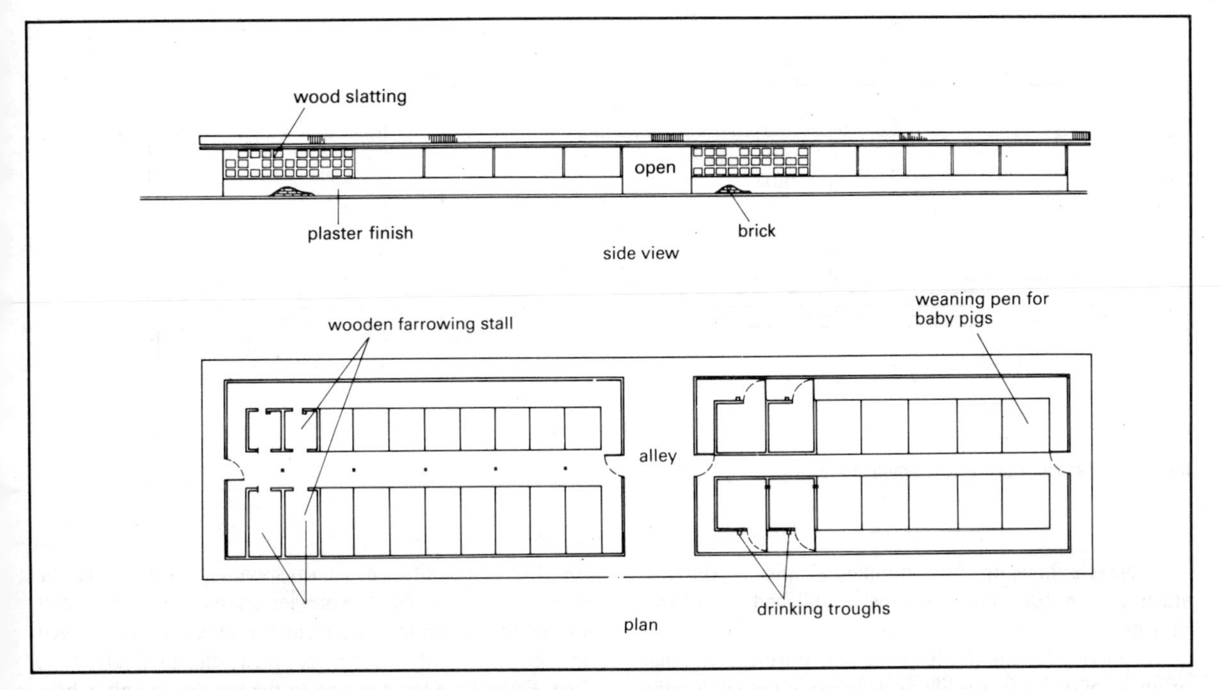

fig. 7.9 Plans for a farrowing, lactating and early-weaning pig house

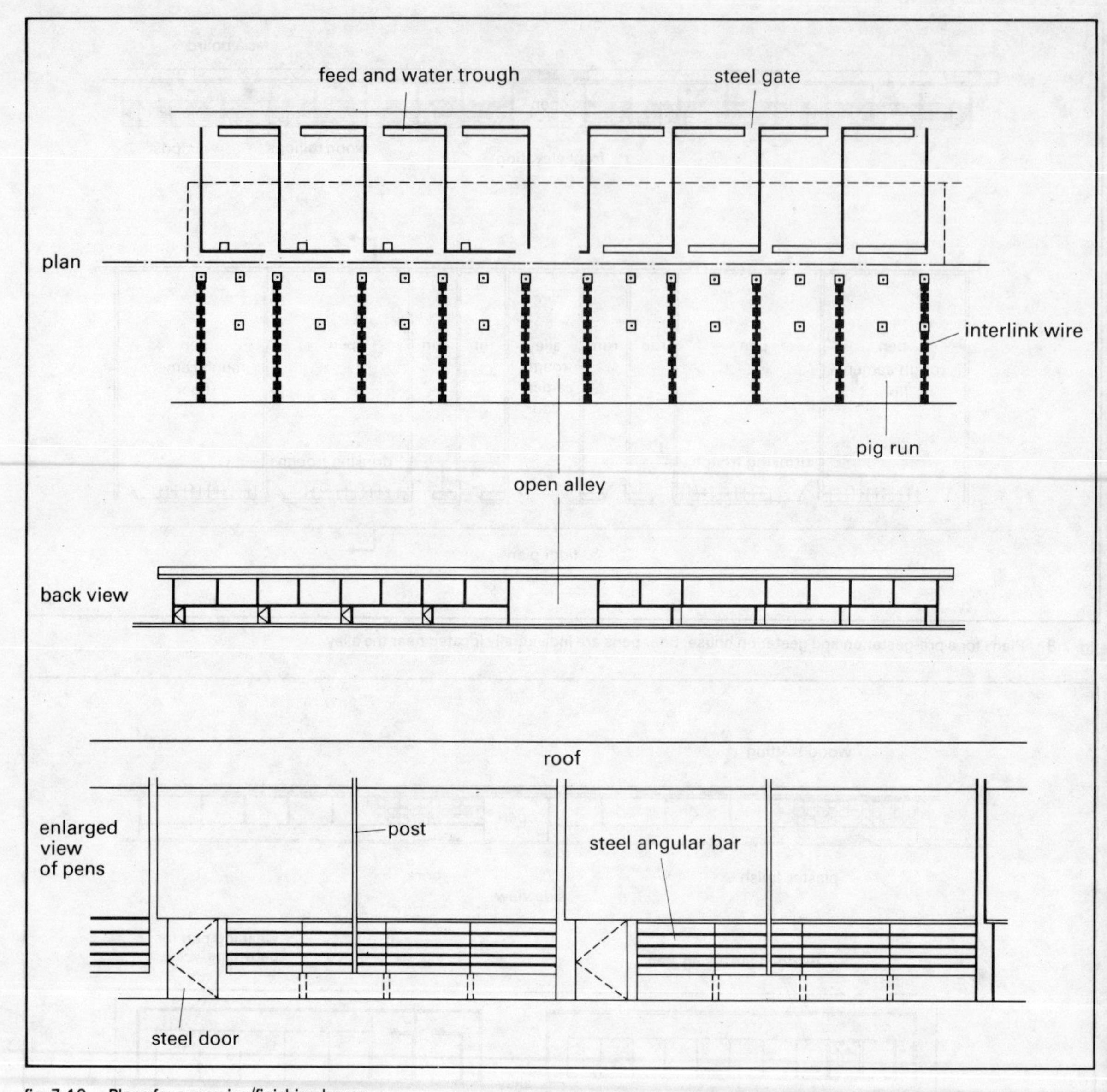

fig. 7.10 Plans for a growing/finishing house

Life-cycle housing for commercial pig production, starting with 100 sows, may be established in an area of 1 ha.

To provide for the needs of the pigs during each stage or phase of their life, the life-cycle housing system is composed of four units, each corresponding to one stage or phase of the pig's life. The pre-gestation and gestation unit houses the breeding sows from the time they are served and fertilised until 3 days before their scheduled farrowing time. From here the sows go to the farrowing unit, where they farrow and stay until weaning time. After weaning, baby pigs are raised in the baby pig nursery unit until they are 8 weeks old, or until they weigh about 12–15 kg. After the baby pig nursery the pigs are housed in the growing/finishing unit until they attain market weight.

Efficiency in managing and handling the animals is achieved by grouping together pigs that have the same feeding and management requirements. For example, the boar pens and the pre-gestation and gestation pen units may be combined in one building and the farrowing and lactating sow unit may be placed in the same building with the baby pig nursery.

It has been mentioned before that what pig raisers look for most in a housing system is the degree to which the pig house can be kept clean and sanitary. In the Tropics especially, conditions in the environment are favourable for the proliferation of disease and parasites. Pig housing, therefore, should be so designed as to prevent or minimise infestation or infection and spread of disease. Pigs in a life-cycle housing system are provided with such protection.

The life-cycle housing system protects the pigs from disease and parasites in a variety of ways. Pigs that may be infected already do not necessarily stay long enough in their pens to enable the parasites or diseases to complete their life cycles. Thus, nursing baby pigs stay with the dams for only 5 weeks, or less, if weaned early. After the baby pigs reach at least 5–6 kg liveweight, usually at 28–35 days of age, they are transferred to the early-weaning nursery pens. At the same time, the dam is brought back to the pre-gestation unit where she waits to be bred again. Thus the baby pigs are totally separated from their mother and there is much less opportunity for the mother's infections to be transferred to her offspring.

The baby pigs stay in the weaning house either only 30–35 days or until they are about 2 months old and weigh 12–15 kg. The pigs that are weaned early are moved to the growing/finishing house where they stay, for another 4–5 months or until they have reached marketable weight.

Gilts that have been selected to replace the breeding stock are kept in the finishing pens when they reach 60–70 kg liveweight. This unit is in the same building. Pens in the finishing section are a little larger than those in the growing section. In these same pens are also kept the barrows that are to be raised to 100 kg. Both the culled gilts and the barrows may be marketed for bacon or they may be butchered as soon as they weigh 60 to 70 kg.

Floor space requirements

The amount of space in the pen required by each animal depends on the size of the animal, the ambient temperature, the ventilation available in the house and the method of feeding. On a slatted floor a little crowding may be tolerated; it may be also practical and economical because the animals' manure can be easily worked through the slots and the pen is much easier to clean.

Table 7.1 shows the minimum allowances of floor space suggested for the different weight categories of pigs kept in close confinement. The floor area for slatted pens applies both when the floor is fully and partially slatted. The extra space required for solid-floored pens is necessary to allow for an adequate standard of cleanliness.

Table 7.1 Recommended minimum floor space allowance per pig for pigs in confinement housing. (*Source:* **Gehbach, G. D., Becker, D. E., Cox, J. L., Harmon, B. G. and Jensen, J. H.** (1966). Effect of floor space allowance and number per groups on performance of growing-finishing swine. *J. Anim. Sci.,* **25**, 386.)

	Floor area (m^2)	
Liveweight (kg)	**Slatted**	**Solid**
11·5 to 18·0	0·27	0·36
19·0 to 45·5	0·36	0·54
46·5 to 68·0	0·54	0·81
69·0 to 95·5	0·72 (0·81 in warm seasons)	1·08
Sows and boars		4–5

Making use of pig wastes

Methane gas production

The production of methane gas from pig manure is made possible by the presence of certain bacteria which do not require air. These bacteria can decompose certain forms of organic compounds into methane and carbon dioxide. This process of decomposition can take place only if:

1 oxygen is excluded from the process;

2 the raw material contains nitrogen;
3 the temperature is favourable – the optimum is about 35°C as a temperature much below or above this retards or even arrests the process;
4 the reaction is slightly alkaline, with a pH of about 7·5.

These requirements mean that the plant or place where methane gas is produced should not have any air inside it, and that the pig wastes or raw materials in it should be protected from too much heat or cold.

A methane gas generator or plant comprises:

1 A tank or digester in which the pig wastes are fermented.
2 A gas storage tank or gas holder in which the gas is collected and stored.
3 A piping system which allows gas to pass from the gas storage through a purifying channel to the outlet in the kitchen or elsewhere.

The gas produced in a digester has a varied composition and only about 53–62 per cent is pure methane. The composition of a typical sample is: 53·8 per cent methane; 44·7 per cent carbon dioxide; 0·3 per cent hydrogen; 0·1 per cent carbon monoxide; 0·1 per cent oxygen and 1·0 per cent other gases.

The gas produced from a simple generator contains about 56 per cent methane and has a calorific value of 25 320 kJ per m^3 of atmospheric pressure (7·56 litres of methane gas at 351 kg per cm^2 is equivalent to 3·78 litres of petroleum).

The gas is non-poisonous, has a very low inflammability and a burning cigarette will not ignite it.

Pure methane burns in air when the mixture is 91 per cent air and 9 per cent methane. Impure methane, as produced by the simple digester, will burn only when mixed with the correct proportion of air; the best ratio is 93 per cent air to 7 per cent impure methane.

Every 40 kg of different types of manure will produce the following quantities of gas:

poultry manure	11·3–12·8 m^3
pig manure	2·0–2·8 m^3
cattle manure	1·1–1·4 m^3

Farmers in Taiwan recommend using sludge from a fish pond or an old sewer, or preferably the fermented manure from a successfully constructed unit as an inoculant to start the process.

A family of three will probably need about 4·2 m^3 of gas daily for cooking and lighting. This amount can be obtained from the manure of 35 pigs.

Fig. 7.11 illustrates an improved design of a methane gas generator which is continuously supplied with raw manure from the pig through a drainage canal. The dimensions are shown in Table 7.2. Details of the digesters are given below.

Digester 1 Concrete hollow blocks or bricks are the most suitable and economical materials for constructing the digester. A square shape is recommended because it is simpler and less expensive to build, although it may be of cylindrical shape.

To prevent water from seeping into the digester, the outer and inner walls should be plastered with waterproofing cement. Both sides of the wall must be plastered simultaneously as the construction progresses. To keep the walls of the gasholder, which are made of plain galvanised iron sheets, from falling on the base of the water seal, the remaining space of about 60 cm from the bottom may be filled with soil. Loss of water due to evaporation may be prevented by maintaining a film of diesel or engine oil on top of the water.

The cement pipe carrying the pig manure to the first digester should be installed so that the spout is 60 cm above the flooring. This provides adequate space for sedimentation and prevents clogging with pig waste. The diameter of this pipe should be approximately 20 cm. The inverted tank which serves as the gasholder should be made of number 16 (15 mm) gauge galvanised iron sheet. In order to reduce heat loss the tank should be painted black after it has been rust-proofed with lead oxide. A hose is connected to the pipe attached to the upper surface of the iron gasholder.

Fermentation and production are shown by the increase and decrease in the height of the gasholder at ground level. When fermentation is poor and production is low, a stirrer should be used. The stirrer consists of 3 or 4 pipes crossed with one another and welded to one end of a 1 m long iron pipe.

When it becomes necessary to clean the inner tank, the gasholder may be removed easily using 4 iron rings welded on the top surface.

Digester 2 The second digester is connected by a second cement pipe to digester 1. Its top opening can be covered with a fixed lid which can be made air-tight by immersing it in a groove filled with water and oil. Two pieces of angle iron are placed on the lid and bolted down to the sides of the walls to keep the lid in place.

Table 7.2 Dimensions of methane generator

Digester 1	Dimensions				
	cross section		length		depth
Inner walls	1·0 m	×	1·0 m	×	2·0 m
Outer walls	1·5 m	×	0·5 m	×	2·0 m
Space between inner and outer walls	7·6 cm				
Gasholder	1·2 m	×	1·2 m		
Floating stirrer	0·9 m	above base of gasholder			
Feeder pipe	20·3 cm	diameter			
Pipe 1	12·7 cm	diameter			
Digester 2					
	cross section		length		depth
Inner walls	1·5 m	×	2·5 m	×	2·0 m
Pipe 2	12·7 cm	diameter			

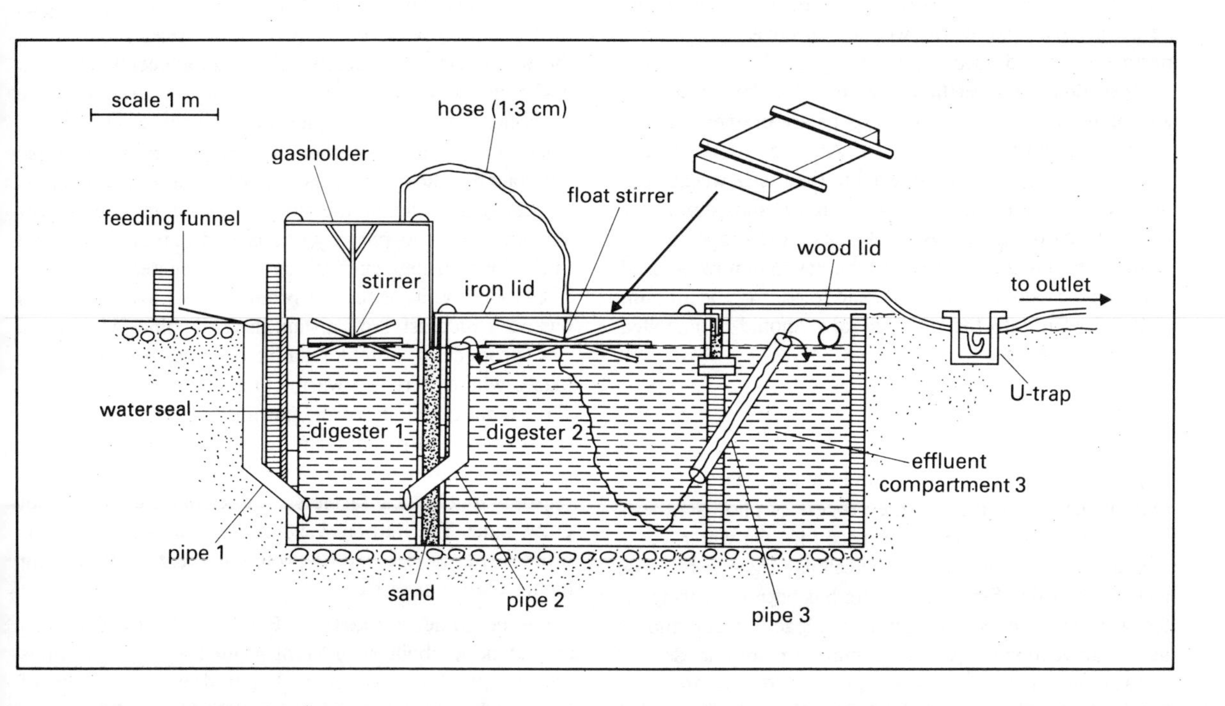

fig. 7.11 Methane gas generator. (*Source:* Chinese-American Joint Commission on Rural Reconstruction, 1965)

Gas produced from the second digester has a much higher percentage of methane than gas made in the first one. This gas in conveyed through a T-pipe and hose to the gasholder in which it is stored. From the other end of the same T-pipe, the gas is led through a PVC pipe to the place where it is to be used. A U-trap made of a PVC T-pipe is fixed at the lowest point to catch any condensation. This U-trap must be about 30 cm high so that the water in it has enough pressure to keep the gas from escaping. It is, however, open at one end to let the condensed water escape.

As shown in Fig. 7.11 a second stirrer is needed for continuous gas production in digester 2. This second stirrer has three floating PVC pipes sealed at both ends. A coarse plastic rope tied to these crossed pipes runs down through an iron ring fixed on the bottom and then upward through cement pipe 2 to the third compartment where the sludge overflows. When this rope is pulled and released it stirs the sludge.

Effluent compartment 3 The third compartment can be of any shape and size, since it only serves as storage for the overflowing sludge. The opening of the third cement pipe should be fixed at the level of the liquid in the three compartments. The end of the pipe, however, should be about 30 cm below the wooden lid covering the compartment. The lid need not be air-tight.

Operation of the methane digester The manure, urine and other effluents from the pig pens are drained into the digester to start the operation. After 5–10 days, gas production begins. Gas produced in the first 2 weeks contains only small amounts of methane and should be released and discarded. After 3 weeks, the proportion of methane begins to increase, and the gas may now be used in the kitchen stove. For safety, and to control the amount of gas flowing into the stove, a valve should be installed along the pipe line.

Algae (*Chlorella* spp.) production from pig waste

One of the by-products of the process of photosynthesis in algae is oxygen which is released into the water.

Cultures of single-celled or unicellular algae were made as early as 1890. Since then, there has been increasing interest in the process. The culture of algae is cheap and its use as an economic feed for animals is recommended.

Type of algae The algal species most common in ponds need not be planted because they can be found growing everywhere. For large-scale culture, the most suitable types are unicellular or non-filamentous members of the Chlorophycea.

It is quite difficult to find a pure culture of one algal species in a pond, but frequently one or two species may predominate. Some of the species commonly observed are *Scenedesmus, Chlorella, Chlorogonium, Chlamydomonas, Euglena* and *Microactinium.* Under normal conditions, *Chlorella* and *Scenedesmus*, either individually or collectively, constitute 95 per cent of the algal population. These species normally adapt to the highly specific conditions of their environment, and it is not necessary to inoculate the pond with other algal species.

Waste as feed for algae In the production of algae to be used for animal feed, organic waste materials are used as the major source of their nutrients. Most algae growing in ponds cannot, however, easily utilise as food nutrients many of the complex organic constituents found in waste materials. So it becomes necessary to culture or grow the algae in symbiosis or partnership with common bacteria. The bacteria break down the organic matter into simple substances that can be readily used by the algae.

It is interesting to understand this symbiotic relationship between the algae and the bacteria. The algae, through the process of photosynthesis, release oxygen in the water. Aerobic bacteria, the actinomycetes and to a negligible extent protozoa, use this oxygen to decompose or break down the complex organic substances in the waste materials to carbon dioxide, ammonia, simple aminoacids, phosphates, and numerous other compounds. The algae assimilate these nutrients and in the presence of sunlight, undergo photosynthesis and manufacture food in the form of new energy-rich cellular materials.

Chlorella culture for the production of livestock feeds has been successful in Taiwan, the Philippines and Japan. The production of chlorella is believed best suited to places where there is a warm climate and ample sunlight. The climatic conditions in Taiwan are somewhat similar to those in the tropical countries of Southeast Asia, Central and South America, and Africa, and there is no reason why the local farmers in these countries should not grow algae. The nutrient yield of chlorella per unit area is 8 times that of rice and its protein content is 50 per cent higher than that of rice.

Culture ponds in practice Fig. 7.12 illustrates a series of four ponds built at different ground levels. The ponds are shallow (approximately 30 cm deep), are made of brick and plastered with waterproof cement. There is a

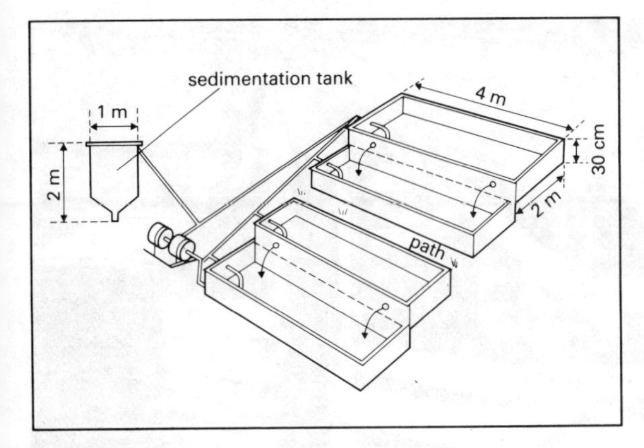

fig. 7.12 Ponds for chlorella culture

vent on the wall adjoining the neighbouring pond constructed at the lower ground level. The bottom near the vent is made concave to permit the accumulation of chlorella during harvest.

In this 4-pond series, 2 ponds can be harvested at the same time by draining the water from them into the adjoining ones. The drained ponds are then refilled with water collected from the lowest pond through a pumping system. The small pump serves three purposes: it pumps up ground water for cultivation, bails out chlorella suspension accumulated in the concave parts of the pond, and refills the upper ponds. No stirring device is provided for this arrangement, although it is desirable that the pond should be agitated several times daily with a scraper.

The pond must be stirred from time to time, so that the algae on the bottom will rise to the surface in order to absorb carbon dioxide. When the algae has grown too thick on the bottom of the culture pond, their further growth is restrained because sunlight cannot penetrate.

Chlorella culture can be associated with methane gas production by using the sludge from the digesters in the pond. Only the fermented manure that is, the fermented manure already in liquid form from the last digester, is recommended as fertiliser for chlorella. However, more benefit can be derived by draining the filtrates off the ferments. The amount of manure to be added daily is estimated at 1 per cent of the cultivation water in the pond. Over-fertilisation causes foaming and should be avoided.

The water should be maintained at pH 7–8 because then the carbon dioxide in the air will dissolve in water to form hydrogencarbonate and become available to the algae. Thus, chlorella cultivated on the farm need not be artificially supplied with carbon dioxide since there is plenty available in the water. If the water becomes too alkaline, the pH can be adjusted to 7–8 by adding lactic acid.

Chlorinated water will inhibit the growth of chlorella. Chlorinated water should therefore be stored at least 3 days before being used as cultivation water in the pond. It is much better to use water from rivers or wells.

The water in the pond should be maintained at a depth of no more than 15 cm so that sunlight can penetrate to the bottom. Chlorella multiplies rapidly only in the presence of sunlight. During rainy days and at night it will settle on the bottom, leaving the surface water clear. The water will turn green again when the sun shines. The algae must therefore be harvested just before daybreak, when it has formed a sediment.

Harvesting chlorella Three to 4 days after starting the algae culture, the pond becomes dark green; it is then time to harvest. The culture water is allowed to stand overnight for sedimentation. Before dawn, the upper layer of the culture solution, which is clear or light green, is drained through the vent into the next pond. The thick chlorella suspension accumulated in the concave part of the pond is scooped out into the sedimentation tank. In this tank, homegrown lactic bacteria that have been cultured in chlorella water are added for fermentation and further sedimentation. The harvested chlorella suspension is allowed to stand for 2 days, after which the upper vent of the tank is opened to drain off the clear water. Finally, the chlorella paste, which is ready for animal feeding, is passed out through the lower vent.

Table 7.3 Digestibility of chlorella in various forms.
(*Source:* **Chung-Po** *(1965). The animal–methane–chlorella cycle.* Chinese, American, Comm. Rural Reconst., Taiwan.)

Form	Percentage digestibility
Normal sun-dried chlorella (powder)	51–54
Crushed chlorella (fine powder)	61–64
Blanched chlorella (dried powder)	70–80
Decoloured chlorella; using alcohol (dried)	73–82
Wet chlorella paste (young cells)	90–92

fig. 7.13 Pig house over a fish pond in Sarawak

Nutritional value of chlorella Workers in California who have been engaged in studies of chlorella have reported that algae have potential value as a feedstuff for pigs. The young cells of chlorella are easy to digest, so the harvested material should not be more than 4 days old. It is also suggested that chlorella should be fed in the raw state as drinking liquid or paste mixed with other feeds in slop form. Dried chlorella is known to be difficult for pigs to digest (Table 7.3).

Pig manure as a fertiliser

Pig manure, like the manure of other animals, may be utilised as a fertiliser. Many pig raisers save it and earn extra income from its sale. It has been reported that 2 kg of manure can be collected daily from a mature pig with an average weight of 150 to 200 kg. In 1 year, a mature pig gives an average yield of 580 to 730 kg of manure. The average chemical composition of pig manure is: 0·70 per cent nitrogen, 0·68 per cent phosphorus, and 0·07 per cent potassium.

In Taiwan, experiments comparing the fertilisation of sugar cane with 20–40 tonnes per hectare of pig manure and with a chemical fertiliser mixture (N, P, K: 120–125, 80–100, 80–100 kg per hectare, respectively) showed no difference in tillering ability, millable cane, and sugar content. It was also found that a combination of chemical fertiliser and 20 tonnes of pig manure per hectare increased cane yield by 15·9 per cent compared with the use of only chemical fertilisers.

Pig manure is also used to fertilise fish ponds (Fig. 7.13). Mud from the fish ponds fertilised with pig effluents may be used to fertilise fruit trees.

Chlorella/fish pond/pig manure/gas recycling system

A recycling system, in an integrated farming operation which involves pig, chlorella algae, aquaculture (pond fish culture), methane gas and vegetable and paddy rice production, has been successfully demonstrated in the University of the Philippines at Los Baños (Fig. 7.14).

In this system, a concrete pond is built on the roof of

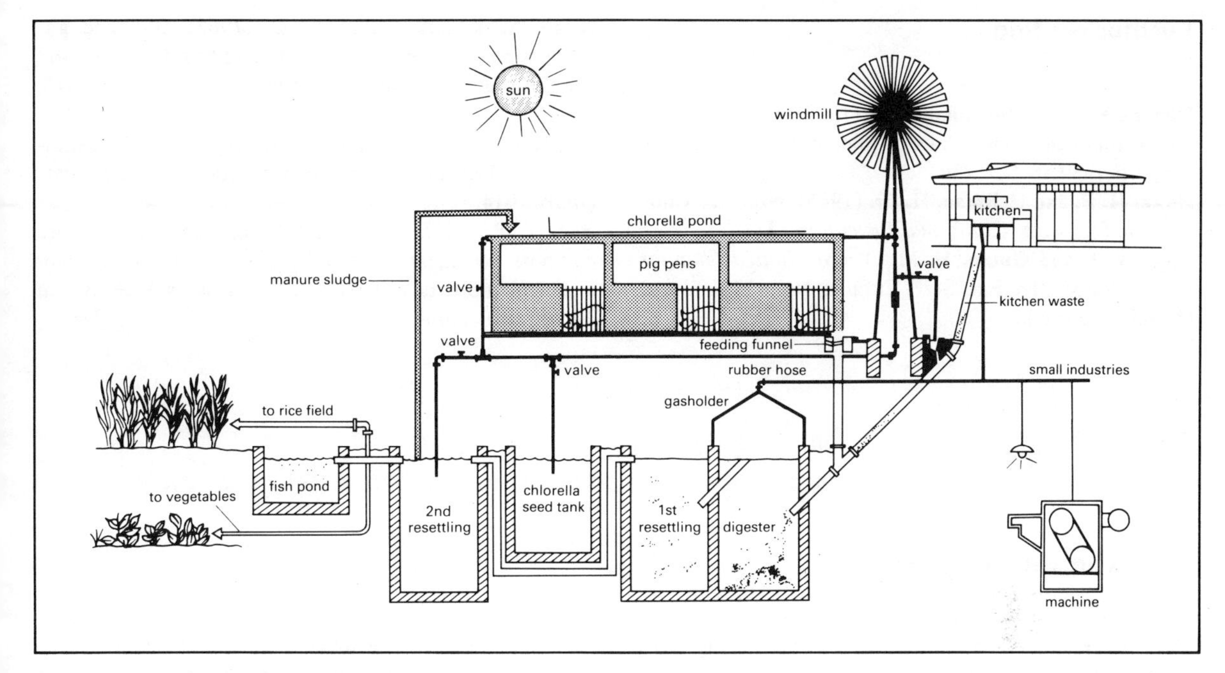

fig. 7.14 A recycling system of agricultural production

the pig house. This pond which is used for chlorella algae culture, provides a cool environment for the pigs, especially during the summer months. Chlorella production is facilitated because of its direct exposure to sunlight.

Harvesting of chlorella is done early in the morning because chlorella settle at the bottom of the pond in the absence of light. The water from the elevated algae pond is always drained into the fish pond to conserve the large volume of water with chlorella washings. These chlorella washings and other algae growing in the pond, serve as feed for the fish (carp or tilapia (*Tilapia mosambica*)). After the algae is harvested by the fish, the water from the fish pond is pumped back to the elevated chlorella pond by the windmill.

When the elevated chlorella pond is sufficiently filled with inoculated algae water, the control valve near the fish pond is closed. The windmill-operated pump then circulates the water in the chlorella pond so that it becomes aerated and chlorella production is facilitated.

The harvested chlorella is used as feed for pigs, either in fresh or in dried form. As protein feed supplement, it is fed to pigs at the rate of 10–15 per cent of the ration.

The pig manure is drained into a methane gas generator. The effluent from the digesters, in which the methane gas is formed, passes into the third compartment where further settling takes place. An outlet in the settling tank provides a slow outflow of clear effluent for fertilising vegetables and paddy rice. Water from the fish pond contains a large amount of algae and is effective as a fertiliser for vegetables.

It is important to point out that neither anaerobic nor aerobic treatment remove sufficient nutrient materials from the manure to make the effluent safe to be discharged into a fish pond. While decomposing organic matter, bacteria will quickly use up all the available oxygen from the water. Lack of oxygen may then suffocate the fish in the pond.

One method of reducing the rich organic matter content of the slurry water is to divert the effluent into a lagoon in which an aquatic plant, such as water hyacinth (*Eichornia crassipes*) or kang kong (*Ipomoea reptans*) is growing. These aquatic plants which normally float on the surface and absorb nutrients from the water can then be utilised as a livestock roughage feed.

Further reading

Hansen, E. L. and Muehling, A. J. (1962). *Hog farrowing houses and equipment.* Univ. of Illinois Ext. Serv. Agric. Home Econ. Cir. No. **780.**
Hintz, H. F. and Heitman, H., Jr. (1965). Nutritive value of algae for swine. *California Agriculture:* USA.
Hug, W. I. and Gutlin, H. M. (1966). *Slotted floor for swine.* Cooperative Ext. Serv. Circular No. 413. Univ. of Hawaii: Honolulu.
Kuan, S. S. and Chour, J. C. (1963). Methane gas production. Anaerobic fermentation of hog droppings and agricultural waste minerals. *Taiwan Sugar Quart.*, **10,** 17–20.
Liu, K. C. (1962). The agricultural review of the sugar industry of Taiwan. VIII. Swine production. *Taiwan Sugar Quart.*, **914.**
Yeh, T. P. (1965). The effect of lime and hog manure compost on sugar cane yield and on certain physio-chemical properties of reddish brown latosol. *Soils and Fertiliser in Taiwan*, p. 110.

8 Record-keeping

Farmers are often faced with the problem of the high cost of producing pigs and the narrow margin of returns from their operations. They cannot, however, provide explanations for their losses. This situation is distressingly common for many pork producers and it often arises from improper and inaccurate record-keeping. Most farmers doubt the value of record-keeping, especially if there is a marked difference between the end-results and their expectations. They also wonder if the benefits derived from record-keeping can compensate them for the time it requires.

Importance of recording

Piggery owners, particularly those raising pigs for breeding purposes, need to keep proper and adequate records. The reliability of pedigree records depends not only on the integrity of the breeder but also on how well the records are kept. Furthermore, intelligent decisions cannot be made without knowing facts. The breeder or producer should collate all essential observations on the farm, and using these, adopt those management practices that will benefit the piggery.

As information is acquired about the piggery based on performance records, each item of fresh knowledge will suggest the next move and often indicate where improvements are possible and necessary. It must be pointed out that no market-orientated herd can succeed or operate at a high level where 'looks' and not performance are evaluated. Physical looks are quite subjective and do not significantly influence the predicted performance of breeding animals.

Records to keep

Records should be simple and easy to understand. Few people can maintain a high degree of accurate recording when a complex set of records is kept. They should be systematically arranged and made to suit the farmer's needs. Some records, like those on breeding, the sow's litter, feed and liveweight, mortality and the monthly inventory records may be altered to meet the varied conditions on the farm.

Breeding record

The format (Table 8.1) gives essential information related to breeding and reproductive efficiency.

1 **The expected date of farrowing** From this, the necessary preparations for the sow's delivery can be easily arranged. The pregnant sow can be isolated from

Table 8.1 Breeding record

Sire		Dam		Date mated	System	Time after onset of heat (hours)	Expected	Farrowed (actual)	Size of litter		Remarks
Herd number	Breed	Herd number	Breed						Male	Female	
12	DJ	18	BS	5.1.71	natural	48	29.4.71	29.4.71	7	4	1 pig dead at birth
14	DJ	16	DJ	3.5.71	natural	24	25.8.71				
19	DJ × LD	28	DJ	3.5.71	natural	24	19.9.71				Recurred 25.5.71

the rest of the breeding herd and placed in comfortable maternity quarters. In this way, by merely examining the breeding record for the expected parturition date of the sow, the litter can be kept safe from crushing by other sows.

As shown in the sample breeding records, a Duroc Jersey (DJ) gilt, herd number 12 mated January 5, 1971, is expected to farrow on 29th April, 1971. The gestation period of a sow is 3 months, 3 weeks and 3 days, or 114 days, as computed from the date of insemination, assuming that she has settled or has had no repeat insemination after 18 to 21 days.

2 **The exact parentage of offspring** This is particularly important where a pedigree record is necessary in the selection of replacement breeding gilts. A pig with a well-kept pedigree record should command a higher price than one sold in market or on farm without records.

Sow's litter record

The sow's litter record reflects the pedigree data. This record shows the relative merits of the parents of each litter. Information on the mating of closely related pigs, such as father to daughter, brother to sister, or son to mother, etc., which should be avoided, is important. Record-keeping for this purpose will prevent the undesirable consequences of such matings.

Table 8.2, which shows litter size and weight at weaning, reflects the mothering ability of the sow. There must be something wrong with a sow which, with a litter of 11 pigs at birth, was able to raise only 5 or 6 pigs to weaning. A careful inspection of her records may reveal the probable cause of her poor performance which may be: defective teats (blind or inverted); poor milking ability; or poor disposition and cannibalism (killing some of her own litter).

If the sow had previously been a good mother, the fault could be due to the stockman. Perhaps the sow was not given the proper feeds in correct proportion during the lactation perid. The care and management given by the stockman can also be reflected in this record when poor performance is shown to be due to deficiencies in the sow.

Feed and liveweight record

Feed and liveweight records (Table 8.3) are very important, especially for those raising pigs intended primarily for slaughter. Feed cost on the average is 75–80 per cent of the total cost of production. It is imperative, therefore, to determine the pig's feed consumption. Accurate feed and weight records show the following points:

1 **Feed efficiency (F/G)** The amount of feed required to produce a 1 kg gain in liveweight is easily computed. For growing/finishing pigs, feed efficiency should be about 4 kg of feed per kilogram liveweight gain.

2 **Gain in liveweight** This is the difference between final and initial liveweights.

Table 8.2 Litter or sow record

Record of dates	Farrowing 28.4.71	Weaning weight taken 28.5.71	6-month weight taken

Information on sow		Information on boar	
Breed Berkshire	Herd number 18	**Breed** Duroc Jersey	Herd number 12
From litter size of:		**From litter size of:**	
at birth	10	at birth	12
at weaning	8	at weaning	8
Date Farrowed:	10.2.69	**Date Farrowed:**	12.1.69
Sow's sire	Herd number 14 (BS)	**Boar's sire**	Herd number 161 (DJ)
Age in months at:			
farrowing 2 years	Sow's 3rd litter (1st, 2nd, etc.)		

Number born living	Male 6	Female 4	Sex unknown 0
Number born dead	Male 1	Female 0	Total number born 11

Pig number	Sex	Birth weight (kg)	Condition at birth (on scale: 4–excellent – 1–inferior)	Weaning weight (kg)	Herd number	Remarks
1	M	0·8	3	5·25	22	
2	M	1·0	4		23	Crushed: 1.5.71
3	M	0·8	3	6·00	24	
4	M	0·6	2		25	Crushed: 3.5.71
5	M	0·9	3	4·75	26	
6	M	0·6	2		27	Sacrificed: 5.5.71
7	M	0·4	0			Born dead: 28.4.71
8	F	0·8	3	4·50	28	
9	F	1·0	3	4·75	29	
10	F	0·7	3	2·25	30	Sacrificed: 28.5.71
11	F	0·8	3	4·00	31	

Table 8.3 Feed and liveweight records

Pen number	Herd number	Sex	Initial weight (kg)	Feed given	Feed consumed	Remarks	Pig's weight in (kg) on date	Feed given	Feed consumed	Remarks

3 **Cost to produce 1 kg gain in liveweight**

It is easy to compute any one of the three items above. For example:

Initial liveweight of pig (at 2 months of age)	10 kg
Final liveweight of pig (at 7 months or for a 5 months feeding period)	70 kg
Cost per kg ration	\$0.10
Total quantity of feed consumed	240 kg

Gain in liveweight = 70 – 10 kg = 60 kg

$$\text{F/G or feed efficiency} = \frac{240 \text{ kg (feeds)}}{60 \text{ kg (liveweight)}} = 4 \text{ kg}$$

Cost to produce 1 kg gain in liveweight
= 4 kg (F/G) × 0·10 = \$0.40

A knowledge of feed conversion data enables a decision to be made as whether to continue feeding the pig the present ration or whether to switch to another feed. It is unnecessary to have a 4- or 5-month feeding period to determine the merits of a specific feed formulation; 2 or 3 months of feeding will be sufficient to evaluate the efficiency of the feed.

Mortality record

The mortality record (Table 8.4) identifies the cause or causes of mortality and indicates ways to reduce the percentage mortality by adjusting management practices. Knowing the cause or the probable causes of death will help minimise, if not totally eliminate, errors in managerial practice. Thus, mortality records indirectly prevent high mortality rates in the herd.

Monthly inventory record

From the data presented in Tables 8.5(a) and (b), one can readily tell whether the business is economically rewarding or not. Total income from sales and/or slaughter of pigs can be easily compared with partial expenses (feeds

Table 8.4 Mortality record

Herd number	Sex	Date of death	Cause	Remarks

Table 8.5 (a) Monthly inventory record (monthly report).

Report of the month of ____________ **19**

A Expenses

1 Feed consumed — Unit price — Cost

______ kg sows ration ______ ______

______ kg starter ration ______ ______

______ kg grower ration ______ ______

______ kg fattening ration ______ ______

______ kg other ______ ______

Total ______

2 Drugs used

______ ______ ______

______

______

Total ______

Total expenses ______

B Income

Livestock and product

1 Sales

Breed	Herd number	Weight (kg)	Official receipt number

2 Slaughter

______

Total ______

Total income ______

C Herd statistics

Number of animals at the begining of the month ______

Increase ______

Decrease ______

Number of animals at the end of the month ______

Remarks

Percentage mortality ______

Date of report **19**

Table 8.5(b) **Reverse side of monthly report**

Livestock inventory for the month of 19
Total number of animals at the beginning of the month
Total number of animals at the end of the month

Class or animal	Breed						
	Landrace	Yorkshire	Berkshire	Duroc-Jersey	Cross	Grade	Total
Young boar							
Stag (castrated boar)							
Sow with litter dry							
Barrow							
Gilts							
Weanlings (less than 5 months) male female							
Total							

consumed and drugs used). If the producer is mixing his own rations the column for feed consumed can be changed to the different feed ingredients (Table 8.6). It may also help to show which inputs of production fail to influence the necessary outputs. The monthly inventory record (Tables 8.5(a) and (b)) also serves as an aid in projecting the size of the herd population for future expansion. Monthly reports may also be collated to compile an annual report.

Supplementary recording

The use of index cards stapled on posts or on the sides of the building, where temporary recording can be done, is recommended. This system makes accurate recording easier because it permits immediate jotting down of observations relevant to each record. The feed record can be stapled on the feed container.

It is very important to spare some time for the analysis of these records. Index cards should be collected weekly or monthly or as often as it is desired and transferred to the permanent records in the main office. Interpretation of the records can easily be done by utilising all this consolidated information. Final analysis of these records may expose shortcomings in management and other faults and result in earlier correction.

Record-keeping is not only beneficial to large- and medium-scale producers, but also to small-scale pig enterprises.

Table 8.6 Monthly inventory record

Report of the month of ______ 19

Feed consumed	Unit price	Cost
______ kg ground maize	______	______
______ kg rice bran (fine)	______	______
______ kg copra meal	______	______
______ kg fish meal	______	______
______ kg soybean oil meal	______	______
______ kg maize gluten	______	______
______ kg skim milk	______	______
______ kg ground shell	______	______
______ kg molasses	______	______
______ kg bone meal	______	______
______ kg vitamin/mineral premix	______	______
______ kg pollards	______	______
______ kg coconut oil	______	______

9 Slaughtering and processing

Successful slaughtering does not improve the quality of pork brought about by breeding, feeding and management, though improper procedures used before and during the slaughtering operation can lower the yield and impair the quality of a pork carcass. Since it is the carcass that finally determines the return on the investment, great efforts should be exerted to make the slaughtering operation successful.

The slaughtering procedures and preparation are essentially the same for all meat animals. Usually the slaughter stock are fasted for 18–24 hours. During this period feed is withheld, although sufficient water is given to the pigs. It has been reported that, based on warm carcass and slaughter weights, unfasted pigs when slaughtered have the same dressing percentage as fasted pigs.

Slaughtering equipment and procedure

The slaughtering materials, knives, scraper, gambrel stick (a curved steel with hooks on both ends for hanging the carcass), cleaver, scalding vat, pork cutting saw and working table should be kept clean and in proper order before the start of the slaughtering operation. The slaughtering procedures are detailed below.

Sticking

This is the term used for bleeding. The animal is first hog-tied and then restrained flat on its side on a concrete table, the lower portion of the neck being exposed upward. The slaughterer holds the head by the left hand and sticks the knife into the hollow portion above the breast bone. A 20 cm knife sharpened at its tip is directed towards the tail, while a thrust and withdrawal is made to cut the jugular veins and arteries. The slaughterer should be careful not to pierce the heart as this may kill the animal instantly and prevent a desirable bleed. The knife must be kept midway between the shoulders to avoid shoulder stick.

Scalding and scraping

Scalding is the loosening of the hair and the scurf; scraping is the removal of these materials from the pig's body. A sufficient amount of water at the correct temperature of 65 to 71°C, certainly not more than 85°C, should be available. Scalding is done by immersing the whole or half of the pig's body in the hot water vat; this is followed by scraping. However, to make the whole process more practical, scalding and scraping are done simultaneously. The head and the legs should be cleaned first, since they are the hardest to scrape. Scalding and scraping of the other parts follow.

Sufficient hot water is poured slowly on the skin and when the hair and scurf slips off easily, scraping can be accomplished. The scraper should be manipulated towards the flow of hair growth to avoid cutting the bristles. The hair and scurf that is not removed is shaved off with a sharp knife.

Removing the head

The head is removed by cutting through the neck at a distance two fingers wide from the base of the ears and severing the connection at the atlas joint.

Evisceration

This is the removal of the visceral organs. The breast is first cut open and the breast bone loosened from its attachment. A long cut is made down the belly from the breast to the hams. In opening the belly, care should be taken not to cut the intestines, by guarding the point of the knife. Then, the pelvic bone is split by following the *pubis symphysis* or the white tissue that separates the hams. Another cut is made down through the diaphragm and abdominal muscles. The intestinal tract and other visceral organs can then be pulled out. While the carcass is still warm, the leaf fat should be removed. The carcass is finally split through the centre of the back bone.

Handling of the entrails

The heart, liver, spleen, kidney and lungs together with the trachea and oesophagus should be separated from the intestinal tract, and the gall bladder should be carefully removed from the liver. The kidney is split open and the whitish tissues inside removed. The stomach, small intestines and large intestines are separated. The stomach fat is taken out; the stomach is opened and flushed with water to clean it. After the mesenteric fat of the intestines has been removed, the intestines are flushed with water and then inverted. Inverting the intestines may be made easier by using a stick. To remove their slime, the intestines and stomach may be macerated with a mixture of salt and papaya (paw-paw) leaves, after which the dirt can be washed out with water. A small amount of vegetable oil can be added before they are macerated further and finally washed again with water.

If the cleaned intestinal tract is intended to be used in cooking, it must be pre-cooked to facilitate chopping. If the intestines are to be used as sausage casings, 5 per cent of salt should be added. They should then be stored for 1 day in a cool room before they are dried.

Chilling

Chilling is the cooling of the carcass at a temperature around freezing. The carcass may be gambrelled after loosening its tendon on both hind legs and then hung as a whole. Each side may also be separated, hooked and hung on the loosened tendons of the hind legs. The carcass should be allowed to hang for about 2 hours before placing it inside the chiller. It should be kept at the chilling temperature of 2·2 to 4·5°C for 24 hours before it is finally cut up.

Cutting a side of pork carcass

To cut a side of pork carcass, knives, a cutting saw and a cutting table are used. The following procedures may be followed.

American system

1 The half carcass is placed on its side on the table with the bone side up and the feet towards the person doing the cutting.

2 Cutting should proceed thus: shoulder – neck bone – jowl – clear plate – fore shank. The third rib is sawn across at right angles to the general line of the back and the shoulder and jowl separated from the rest of the carcass. The fat is removed from the upper two-thirds of the shoulder, leaving only about 0·33 cm. The neck bones are detached. The shoulder is turned over with the skin side up and the jowl removed by cutting at the junction of the shoulder. The Boston butt should not be separated from the picnic, but left as skinned shoulder, to be separated later. The fore shank should be removed by sawing across the forelegs just above the knee. The jowl is cut up into fat and lean trimmings.

3 The ham is sawn across the junction of the 3rd and 4th sacral vertebrae, about 6 cm in front of the pubic bone. This cut should be perpendicular to a line bisecting the hind foot and the hock. It should be sawn only through the bone and a knife should be used to cut the lean. The tail bones are removed. If the ham is to be skinned, the skin and fat should be removed from the upper two-thirds of the ham. Only 0·33 cm covering of fat should be left over the skinned portion of the ham.

4 The loin and backfat should be separated from the side by sawing the ribs from a point as close to the backbone (chine) as possible at the shoulder end to a point just below the tenderloin (psoas) muscle at the last rib. Then a cut should be made on a curved line from the lower edge of the tenderloin muscle, at the ham end of the loin, to the point where the last rib was cut by the saw. From this point forward the cut should follow the division in the ribs made by the saw.

5 The spare ribs should be lifted out by cutting as close to the bone as possible. A small amount of rib cartilage should be left on the side instead of cutting deep enough to get it out entirely. After the flank has been worked out square, the side is trimmed as fully as possible on the underline and on the flank end. It is not necessary to remove all of the teat-line, and no trimming should be required on the shoulder end or on the loin side.

6 The trimmings should be separated into fat, lean and skin, and the amount of fat put in the lean trimmings should not exceed 25 per cent. Figs 9.1 and 9.2 show the different cuts on a chilled pig carcass.

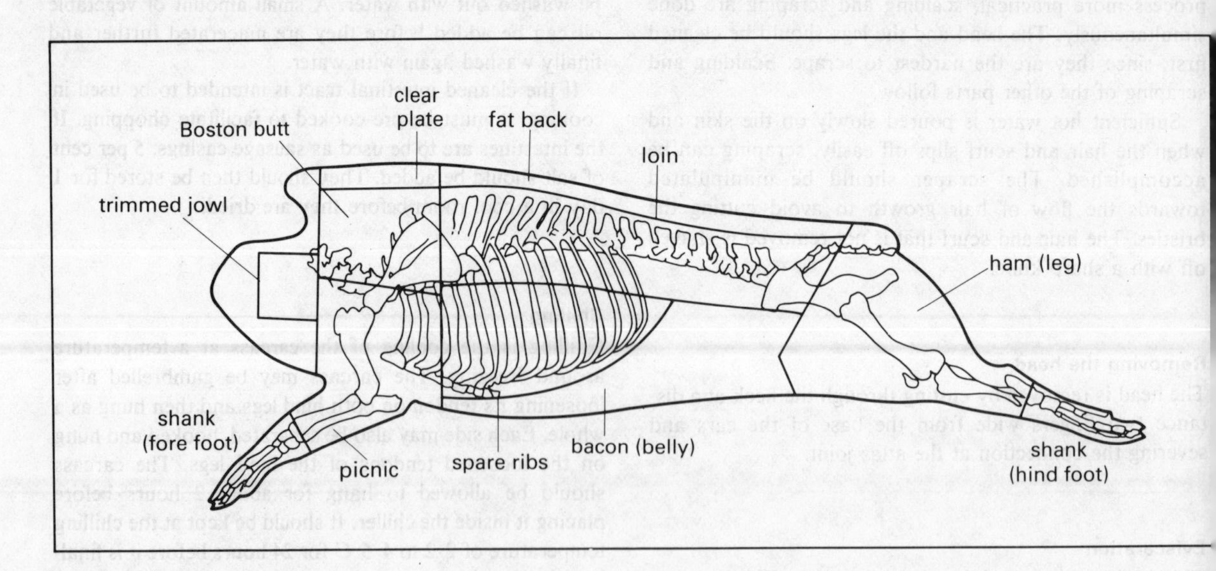

fig. 9.1 Wholesale cuts of pork and their bone structure

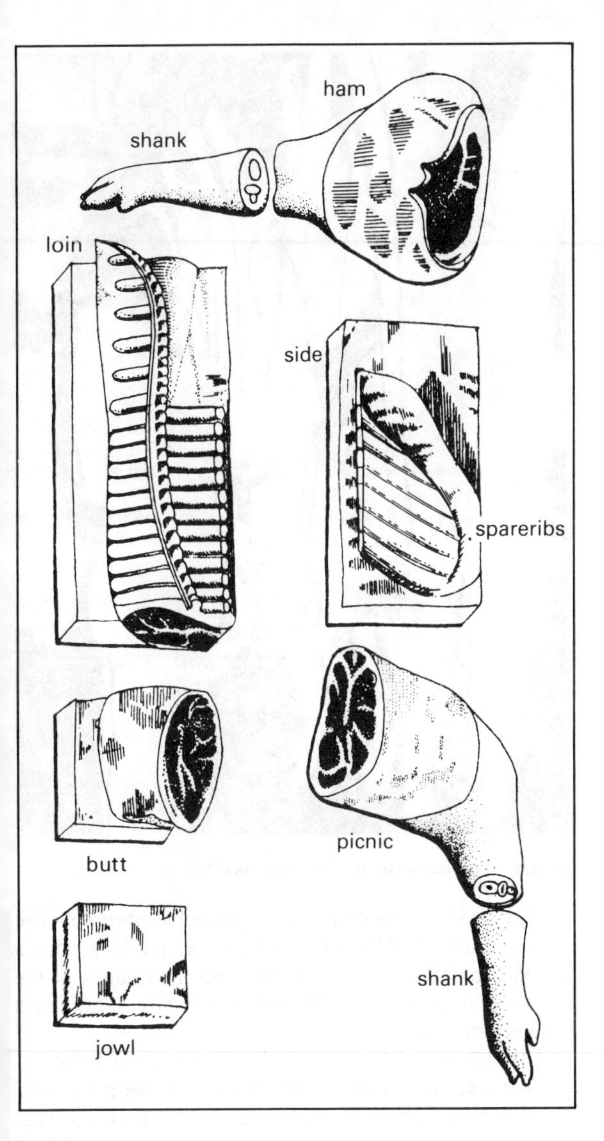

fig. 9.2 Pork cuts

British system

The method of cutting varies considerably in different parts of the UK. In southern England and Denmark (a major exporter to England) the pigs are utilised for bacon by adopting the Wiltshire method, while in the English midlands, Northern Ireland and Ayrshire (Scotland), the pigs are used for pork, and for manufacturing purposes.

In the preparation of the sides for curing, using the Wiltshire method, the secondary offals, such as head, feet, tail, backbone (chine), tenderloin, aitch bone (pelvic bone), blade bone (shoulder blade), kidney or flare fat, are all removed. This leaves the side or 'flitch' to be cured in one piece. The hind quarters are cut off after curing and this is known as the gammon. The fore-end is cut into several pieces including picnic ham from the foreleg. The backfat is removed from the middle and is sometimes cured and sold as backfat. The lean meat of the loin is left with very little fat and provides a high quality type of lean meat cut. The rasher bacon, left after removing these cuts, is not of as high quality as the back bacon from Wiltshire sides, which includes the 'eye' of lean meat from the back.

In Northern Ireland, the preparation of the carcass for pork differs from that of bacon. Usually, upon arrival at the factory, the head is first cut off and hung up. The carcass is then cut straight down the back and the backbones are taken out. The ribs and bone of the fore and hind legs are removed or sliced off with a fair quantity of meat attached to them. These are a delicacy which are stewed or roasted, but if too much meat is taken the resultant bacon will be over-fat. The hams are cut off and cured separately and the shoulders removed for separate curing. The skin is left on and after curing, the side is rolled and tied up with thick strings in rings 2·5 cm apart. The bone hams and shoulder are also tied with string.

In Ayrshire, the preparation of the carcass for pork is much the same as in Northern Ireland, except that upon arrival at the factory, it is skinned, and the curing is milder than that practised in Northern Ireland. The boned sides are also rolled and tied with strings, but are normally sold unsmoked.

In the English midlands, although the carcass for pork is prepared in the same way as in Northern Ireland and Ayrshire, the pigs are classified into two categories: lean-backed and fat-backed.

1 **Lean-backed side** The whole of the side is cured into bacon. This is a special cut side, which is identified with Wiltshire side after the ham and forehock are removed. The ham is cured separately. The backbone and tail piece are sold fresh. The head, trotters and hocks, although sometimes sold fresh, are usually pickled for the production of cooked meats.

2 **Fat-backed side** The ham, head, forehock, feet, tail and breast bone are removed, and the loin is taken out. The shoulder belly is left for curing. The back fat of the

loin is removed and used to produce lard. The loin is sold as fresh pork and the rest of the carcass is disposed of in a way similar to that of the lean-backed pig.

The kidneys remain in the pork carcass and are sold as part of the loin cuts. The leaf or flare fat around the kidneys and the lining of the abdomen are removed as secondary offals in trimming a carcass for curing.

The Meat Laboratory Section of the Department of Animal Husbandry, University of the Philippines College of Agriculture, studied the proportion of the different wholesale cuts of chilled pork carcasses (Table 9.1). This information should serve as a guide to butchers on what is the expected cut meat recovery from a given weight of finished pig. In this case, the dressing percentage was assumed to be 70 per cent.

Table 9.1 Carcass yield of Landrace × Yorkshire × Philippine pigs (*Source:* **Argañosa, V. G., Alcantara, P. F., Icasas, R. A. and Rigor, E. M.** (1970). *Meat and animals carcass evaluation. 1. Relative proportion and values of the different wholesale pork cuts.* Phil. Soc. Anim. Sci., 7th Ann. Convention, Manila.

Cuts	Percentage of chilled carcass
Trimmed weight	
Shoulder	16·0
Loin	16·7
Ham	19·2
Belly	16·4
Spare ribs	3·2
Neckbone	1·9
Feet	3·0
Fat	19·7
Lean trimmings, Jowl	3·4
Total	99·5

Uses of slaughter by-products

The slaughter by-products of pigs are rather difficult to define. The terms by-products and offal are generally used for those parts or portions which are not included in the dressed carcass. However, the by-products and offal are further divided into edible and non-edible portions. The classification of edible parts of slaughter pig by-products depends on the religion, customs and food habits of the people in the country. Fig. 9.3 shows pork carcasses hung in a meat factory in Zambia.

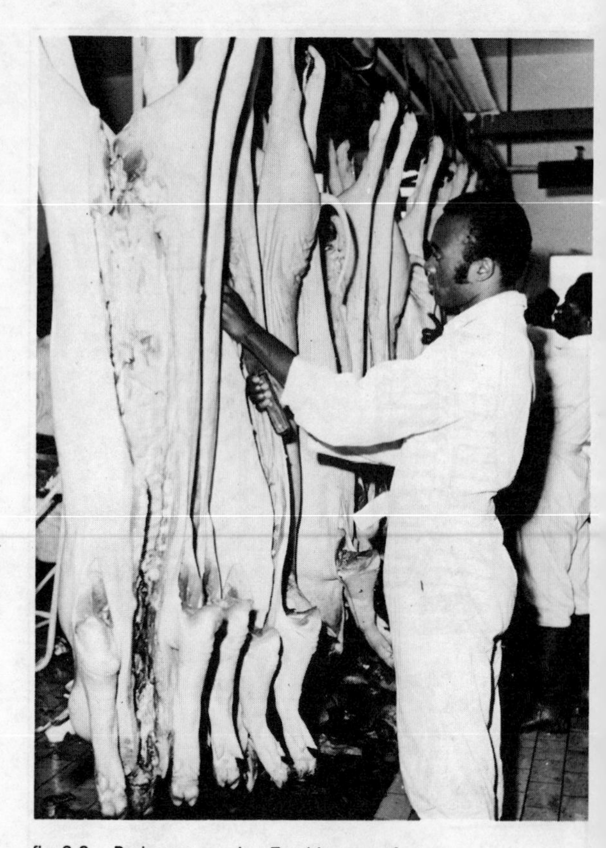

fig. 9.3 Pork carcasses in a Zambian meat factory

The kidneys, heart, brain, liver, tongue and the intestinal tract are generally classified as edible parts. The small intestines are also used as casings for pork sausage. In some countries they are used for the manufacture of guitar strings. Other organs, such as testicles, spleen, lungs, uterus and the blood are eaten by some people in some countries.

Non-edible by-products such as the teeth, bones, hoofs, hair, bristles and glands are used in the manufacture of industrial products. Offal which includes the skin, bone, blood and hoofs are used in a wide range of manufactured items, such as leather, bone meal, blood meal and glue. Pig's blood is used in the manufacture of peptone or lecithin which are obtained from the fibrin. Sometimes, blood albumin is used as substitute for egg white in ice cream making or in bakeries and in the manufacture of waterproof glue. In England, the blood can be used to make black pudding. It is gently stirred as it is collected from the animal to prevent coagulation. It is important that fresh blood be used and these puddings should be made on the same day that the animal is slaughtered.

In Latin America, particularly in Colombia, the blood is mixed with ground pork for making a sausage called *morcia.* In the Philippines the blood is cooked with the liver, spleen and small intestines, to produce a delicacy known as *dinuguan.*

In the Philippines and Hawaii all the hair is scraped off the head. After cleaning the ears and snout, the whole head is roasted. This is called *lechon*, and is a special delicacy at many gatherings and festivities.

Another delicacy made from the head in the Philippines is called *kilawin.* After removing the eyes, nasal passages and teeth and after cleaning the ears and the snout, the head is boiled for 30 to 45 minutes. Then it is chopped into small pieces and mixed with vinegar, sliced onions and ginger or green pepper.

The glands may be used in the manufacture of pharmaceutical products. For instance, the testicles or ovaries are sources of oestrogen, and the pancreas of insulin. Other hormones may be extracted from the pituitary and thyroid glands.

Pigs that die or the condemned parts of the animals in slaughter houses can be processed as meat meal for use in concentrate animal feeds.

Further reading

Davidson, H. R. (1966). *The production and marketing of pigs.* Longman: London.

Heinrickson, R. L. and Gillis, V. A. (1968). *Meat Technology Manual.* Oklahoma State Univ.: Stillwaters: Oklahoma.

Mann, T. (1962). *Processing and utilisation of animal by-products.* FAO: Rome.

Ziegler, T. P. (1958). *The meat we eat.* 5th ed. Interstate: Danville, Illinois.

Appendix

Approximate composition of some tropical feedstuffs

Name	Expressed as percentages (dry matter + moisture = 100%)					
	Dry matter	Crude protein	Ether extract	Crude fibre	Ash	Nitrogen-free extract
Concentrates						
Acacia (raintree), *Samanea saman*						
pods with seeds	89·25	12·79	2·98	14·54	3·27	55·67
pods without seeds	81·51	9·64	1·26	9·43	4·01	57·17
Anchovy, *Stolephorus commersoni*	78·84	42·21	4·58	7·41	16·74	10·98
Blood meal	88·89	76·67	1·08	1·19	3·99	3·83
Bone meal	91·63	26·03	2·66	2·26	53·05	9·33
Brewer's spent grains	89·71	15·40	3·80	18·05	3·76	48·70
Brewer's yeast	93·10	48·80	0·40	4·20	9·40	30·30
Cassava (tapioca, manihot, yuca), *Manihot esculenta*	89·11	4·82			0·10	84·19
Coconut, *Cocos nucifera*						
desiccated	95·44	7·00	64·59	3·45	1·68	18·76
coconut meat						
fresh	35·41	3·35	19·42		0·97	
dried	89·00	7·58	69·09			
coconut extract residue (fresh)	90·80	6·10	5·00	34·70	1·50	24·30
copra cake	90·28	17·53	8·75	12·19	6·12	46·63
copra meal						
expeller process	87·17	17·50	5·77		7·26	56·64
solvent process	89·25	20·57	3·95	12·46	7·14	45·13
paring cake	89·69	17·30	6·08	12·81	5·97	48·70
paring cake, sunripe	76·69	14·25	10·11	12·33	4·11	38·89
paring meal	84·24	17·81	8·94	12·55	4·75	41·48
Cowpea, *Vigna unguiculata*						
raw	87·06	23·50	1·06	5·72	3·34	53·44

Approximate composition of some tropical feedstuffs—*(Cont'd)*

Name	Expressed as percentages (dry matter + moisture = 100%)					
	Dry matter	Crude protein	Ether extract	Crude fibre	Ash	Nitrogen-free extract
heated	89·53	24·80	1·57	4·00	3·77	55·39
Fish meal, species not identified	89·44	53·44	4·39	1·74	21·51	9·38
Goby, *Glossogobius giurus*	88·04	56·64	4·50	2·56	25·71	
Jack (sword) bean, *Canavalia ensiformis*						
raw	92·03	25·75	1·82	9·80	3·73	50·95
roasted	94·14	21·69	2·87	12·16	3·96	53·46
Job's tears, *Coix lachryma*						
bran, coarse	90·47	3·51	2·11	31·89	14·80	38·87
bran, fine	90·25	5·89	5·28	25·63	17·87	35·60
germ	89·04	15·04	2·75	5·66	14·41	51·18
grain (white portion)	89·98	11·57		0·28	0·18	76·95
Maize (corn), *Zea mays*						
maize and cob meal	90·53	10·19	8·70	6·23	1·87	63·54
bran						
coarse	100·00	12·31	7·50	7·84	5·00	67·36
coarse, white	85·89	10·64	7·41	5·51	4·77	57·57
coarse, yellow	84·18	10·68	7·26	5·43	4·60	55·84
fine	100·00	10·89	5·28	3·92	3·15	76·80
fine, white	85·20	10·57	7·26	5·07	3·93	57·76
fine, yellow	88·12	8·25	2·91	1·86	2·02	73·69
cake	90·64	14·65	3·95	6·30	6·25	59·49
cornstarch	89·79	7·28			0·90	82·42
endosperm	100·00	8·73	2·93	2·67	1·39	84·25
germ meal	92·91	14·89	10·50	8·55	3·66	55·31
gluten feed	91·00	24·66	2·17	5·06	1·80	57·31
gluten meal	91·00	54·97	2·30	4·00	2·40	39·50
grain						
high moisture	70·00	7·91	1·89	2·38	1·22	56·60
opaque-2, dent	87·63	8·27			1·53	
opaque-2, flint, white	86·04	8·18			1·54	
opaque-2, flint, yellow	91·35	7·43	1·53	2·76	1·92	77·71
white	86·12	8·90	4·47	2·80	2·61	67·28
white, immature	32·90	4·40	0·80	2·10	0·70	24·90
yellow, flint	82·90	9·00	2·40	2·30	1·60	67·60
yellow, cracked	88·19	8·76	2·87	2·22	1·59	72·75

Approximate composition of some tropical feedstuffs—(*Cont'd*)

Name	Dry matter	Crude protein	Ether extract	Crude fibre	Ash	Nitrogen-free extract
	Expressed as percentages (dry matter + moisture = 100%)					
grits						
yellow	87·96	7·71	0·75	2·03	1·46	76·02
white, No. 10	86·77	9·54	1·43	3·20	0·61	72·00
white, No. 12	86·63	9·48	1·17	0·73	0·93	74·32
white, No. 14	86·64	8·27	0·22	0·76	0·56	76·83
white, No. 22	86·44	8·10	2·05	1·44	1·07	73·11
Rhodesian, No. 10	87·39	11·19	3·50	1·20	0·83	70·67
Rhodesian, No. 12	87·03	9·61	1·30	1·00	0·85	74·27
hulls	100·00	12·28	7·32	9·61	3·98	67·57
hulls/bran mixture white	85·76	8·40	3·10	3·93	1·55	68·78
hulls/grits mixture white	86·48	8·89	2·61	5·11	1·65	68·23
meal						
white	91·90	8·70	1·10	1·40	0·30	80·40
yellow	95·40	8·80	4·30	0·10	1·40	80·90
middlings						
white	85·40	7·49	0·79	1·90	1·04	74·47
white, No. 10	85·76	7·68	0·33	2·05	0·55	75·32
white, No. 12	89·46	7·66		1·38	0·44	79·99
white, No. 14	85·58	7·31		1·65	0·38	76·26
white, No. 16	85·32	7·80	0·29	1·77	0·57	75·08
Meat and bone meal	92·01	45·92	9·50	1·57	30·74	6·68
Meat meal	93·71	66·02	12·94	1·70	2·74	10·31
Molasses						
bagasse	78·38	2·77	0·59	7·97	5·58	61·47
cane	79·15	0·81	1·81		7·27	69·27
pith	82·41	6·75	0·58	16·81	10·66	47·61
black strap	80·85	20·19	1·34	2·36	12·34	44·62
Mungo (green gram), *Phaseolus aureus*						
raw						
yellow	88·39	23·11	1·12		3·44	
green	90·00	19·77	1·23		3·48	
red	86·91	16·70	0·68		3·01	
variety not specified	88·90	22·66	1·31	5·35	3·43	55·43
heated						
autoclaved for 30 min (15 psi), dried	90·75	23·88	0·92	6·07	3·52	56·36
boiled for 25–30 min, dried	86·21	23·85	0·70	4·71	3·42	53·53
roasted for 30 min	92·70	24·45	0·96	7·50	3·10	56·69
meal	85·10	24·84	1·54	5·93	4·23	48·56

Approximate composition of some tropical feedstuffs—(*Cont'd*)

	Expressed as percentages (dry matter + moisture = 100%)					
Name	**Dry matter**	**Crude protein**	**Ether extract**	**Crude fibre**	**Ash**	**Nitrogen-free extract**
Oats (Avena)	90·90	5·81	2·04	12·74	4·45	65·86
Peanut oil meal, *Arachis hypogaea*	91·52	43·65	5·87	5·01	3·80	33·20
Pigeon pea, *Cajanus cajan*						
raw	87·11	20·46	0·88	6·09	3·97	55·72
heated	88·42	20·84	1·32	6·37	4·21	55·69
Rice, *Oryza sativa*						
rice bran (fine)						
1st class	86·59	11·64	11·93	7·20	8·87	46·85
2nd class	87·23	8·11	3·57	18·77	15·70	40·88
3rd class	87·97	7·82	3·32	20·94	16·19	39·70
rice bran (coarse)						
1st class	85·71	7·50	3·80	21·77	14·70	38·13
rice bran						
coarse	86·00	6·61	3·64	21·83	15·52	39·79
fine	86·19	11·56	8·41		15·48	50·74
mixture of fine and coarse	89·17	5·96	1·98	24·97	15·63	40·51
rice bran/hull mixture	84·39	7·93	4·27	15·71	12·72	43·92
rice grain						
broken	86·25	7·44	0·19	1·48	0·61	75·85
cooked	28·12	1·92				
dehulled	86·18	7·46	0·88	1·51	0·74	75·08
polished	88·38	7·73			3·86	76·79
rough	88·53	7·37	1·76	10·44	7·54	62·20
rice meal	90·04	13·45	14·80	11·20	8·37	42·22
rice middlings	100·00	10·11	3·42	4·47	3·13	78·87
fine	85·29	9·69	2·66	2·97	3·33	66·73
coarse	85·54	8·01	0·93	1·18	0·90	74·52
rice middlings/bran mixture, fine	83·55	10·39	5·35	8·54	8·12	51·16
rice middlings/hull mixture						
fine	84·87	7·23	4·77	15·86	12·19	44·84
coarse	87·75	5·07	0·13	22·86	11·08	48·61
rice middlings with rice hulls						
fine	86·76	8·13	1·58	6·13	4·26	66·65
coarse	88·15	6·62	1·37	3·32	2·64	74·20

Approximate composition of some tropical feedstuffs—*(Cont'd)*

Name	Dry matter	Crude protein	Ether extract	Crude fibre	Ash	Nitrogen-free extract
	Expressed as percentages (dry matter + moisture = 100%)					
Rice bean						
Phaseolus calcaratus	89·90	18·40	3·10	7·30	3·90	64·50
with hulls						
raw	88·01	17·42	2·31	7·25	4·02	56·48
heated	88·20	17·64	0·70	6·63	4·22	60·05
Roundscad						
Decapterus spp.	87·53	50·72	4·04	0·82	27·17	4·79
shark	88·01	64·23	2·89	0·91	14·24	1·64
Peruvian	90·12	61·21	2·84	0·88	13·60	10·93
Panama	90·95	74·20	3·69		17·99	
Sea snake						
dried without salt	88·72	58·76	5·93		14·95	9·08
Shrimp						
Penaeus spp.						
meal	85·45	55·61	2·61		19·73	7·50
Skim milk						
liquid 1:9	9·35	2·65			0·60	6·10
powder	89·44	35·48	1·52	1·56	7·51	43·23
Snails,						
African,						
Achatina fulica						
meat, dried	89·91	45·91	8·57		7·75	27·68
Paludina angularis						
meat and shell, fresh	29·30	3·50	0·20	0·30	2·90	22·40
meat and shell, dried	98·04	11·70	2·54		71·22	15·02
Sorghum,						
Sorghum vulgare						
fines	88·62	5·40	2·66	4·43	3·64	72·51
grain	87·92	9·29	2·54	2·46	1·59	72·09
Soybean,						
Glycine max						
bean	90·64	31·38	17·74		4·39	
oil meal	88·37	43·70	1·53	6·42	6·28	30·32
grits	83·57	45·82	0·82	2·66	5·80	28·47
Sugar, brown	93·11	0·16			0·46	92·49

Approximate composition of some tropical feedstuffs—*(Cont'd)*

Name	Dry matter	Crude protein	Ether extract	Crude fibre	Ash	Nitrogen-free extract
	Expressed as percentages (dry matter + moisture = 100%)					
Sunflower seed, *Helianthus annus*						
with seed coat	91·21	25·62	23·32	22·25	4·35	15·76
without seed coat	89·99	31·16	39·79	5·96	9·83	3·25
Therapon ayungin, *Therapon plumbeus*						
dried	85·00	56·00	3·30	0·50	6·20	19·00
dried, salted	97·74	58·72	14·05	1·97	29·00	
dried, not salted	93·14	58·72	20·00	2·67	23·10	
Tilapia, *Tilapia mosambica*	95·80	46·72	11·07	1·79	23·91	12·31
Velvet bean, with hulls *Stizolobium deeringianum*	88·98	24·02	2·58	4·49	3·93	53·96
Wheat, *Triticum sativum*						
bran						
hard	89·14	15·44	3·11	10·07	4·90	55·61
soft	88·65	11·32	2·12	7·53	3·90	63·78
Bulgur	87·52	12·31	0·60	1·40	2·28	70·69
feed	87·95	12·09	1·03	3·13	2·22	69·48
flour	87·54	12·51	0·45	0·34	0·75	70·58
germ	86·82	24·61	9·26	5·15	4·43	43·37
mill feed 'D'	89·60	17·93	5·29	6·67	4·10	44·62
pollard						
hard	88·18	17·15	3·79	8·67	5·15	53·69
soft	88·32	12·95	4·51	7·38	5·98	58·22
rolled	86·04	14·54		3·16	2·65	
shorts	91·17	15·44	4·89	7·25	4·23	59·36
Yeast						
coconut	88·01	31·64	34·61	10·10	5·60	6·06
from coconut water	94·53	31·18	26·58	7·08	7·58	21·45
from sugar cane alcohol distillery	88·19	38·76	1·17	4·95	7·97	35·32

Approximate composition of some tropical feedstuffs—*(Cont'd)*

Name	Dry matter	Crude protein	Ether extract	Crude fibre	Ash	Nitrogen-free extract
	Expressed as percentages (dry matter + moisture = 100%)					
Tubers, fruits and fruit by-products						
Banana,						
Musa sapientum						
flour, from green banana rejects	96·80	8·50	2·10		8·90	
fruit with peelings						
mixed varieties, raw, dried	89·56	5·35	1·07	3·52	5·14	74·48
plantain, green, mature, cooked	91·13	3·17	3·26	4·63	4·09	75·99
plantain, green, mature, uncooked	90·77	3·29	3·10	5·24	4·38	74·76
plantain, ripened with carbide, uncooked	80·47	3·07	3·85	5·28	4·20	64·08
Cashew,						
Anacardium occidentale						
juice extracted 13 h after picking	85·53	8·62	5·86	7·93	2·71	60·41
juice extracted immediately	83·47	8·61	7·30	7·48	2·31	57·77
juice not extracted	86·90	7·76	3·90	6·65	3·62	64·97
Cassava,						
Manihot esculenta						
with peelings, fresh						
white, 4 months	24·51	2·09	0·66	1·98	0·77	19·01
white, 6 months	29·96	1·93				
yellow	36·60	0·84	0·41	1·29	0·94	33·12
with peelings, dried						
white	95·60	2·55	0·80			
yellow	94·48	2·43	0·96	3·87	3·15	84·27
Elephant yam,						
Amorphophallus campanulatus						
raw, fresh	27·10	1·61	0·30	0·46	0·63	12·71
raw, dried	91·50	5·36				
cooked, dried	94·80	7·39	1·15	5·49	5·93	74·84
Irish potato,						
Solanum tuberosum						
raw, with skin	19·26	2·61	0·08	0·45	1·02	15·11
raw, without skin	19·30	2·40	0·10	0·40	0·80	16·00
Jack fruit,						
Artocarpus heterophyllus						
edible portion, raw	14·80	2·00	0·60	2·60	0·70	11·50

Approximate composition of some tropical feedstuffs—(*Cont'd*)

Name	Expressed as percentages (dry matter + moisture = 100%)					
	Dry matter	Crude protein	Ether extract	Crude fibre	Ash	Nitrogen-free extract
Mango,						
Mangifera indica						
Cambodian variety	88·43	14·25	3·14	33·17	10·94	26·93
peelings, Philippine variety	84·89	1·89	1·56	11·76	6·70	62·98
Muskmelon,						
Cucumis melo	86·38	8·59	1·36	23·39	11·30	41·74
Papaya,						
Carica papaya						
fruit, green	6·80	1·00	0·10	0·80	0·50	5·20
fruit, green dried before analysis	93·37	10·59	1·95	13·69	8·65	58·47
Pineapple,						
Ananas comosus						
wastes (rind, pith) of						
Philippine variety	11·70	0·42	0·14	1·39	0·61	9·14
Sweet potato,						
Ipomoea batatas						
tuber, with skin, raw, fresh						
orange variety	21·49	1·21				
yellow	26·72	0·95	0·57	1·55		
white	28·80	0·60	0·40	0·60	1·00	26·80
tuber meal						
purple variety	91·01	3·08	1·06		2·28	
red variety, without skin	95·08	1·86	1·42		2·79	
white variety	92·52	3·53	2·37			
yellow variety	89·71	2·00	0·57		2·62	
meal, roasted	86·25	2·25	1·47	4·31	3·40	74·82
Taro,						
Colocasia esculenta						
tuber with peelings, fresh	25·55	1·46	0·52	0·92	1·15	21·51
tuber with peelings, dried	93·15	5·36	1·80	3·43	4·23	78·34
Yam,						
Dioscorea alata						
tuber purple, fresh	25·00	1·98	0·28	1·43	1·05	20·26
tuber meal	90·62	7·18	1·03	5·17	3·80	73·44
Yam bean,						
Pachyrrhizus erosus						
peeled, fresh	10·50	1·10	0·20	0·50	0·30	8·90
unpeeled	89·60	6·31	2·35	7·20	4·41	69·33
Tukal (aquatic tuber)	46·05	5·75	0·14		1·52	

Index